Traffic Flow in Switching Systems

The Artech House Telecommunication Library

Telecommunications: An Interdisciplinary Text, Leonard Lewin, ed.
Telecommunications in the U.S.: Trends and Policies, Leonard Lewin, ed.
The ITU in a Changing World by George A. Codding, Jr. and Anthony M. Rutkowski
Integrated Services Digital Networks by Anthony M. Rutkowski
The ISDN Workshop: INTUG Proceedings, G. Russell Pipe, ed.
Telecommunication Systems by Pierre-Gerard Fontolliet
Introduction to Telephone and Telephone Systems by A. Michael Noll
The Complete Guide to Buying a Telephone System by Paul Daubitz
The Deregulation of International Telecommunications by Ronald S. Eward
The Competition for Markets in International Telecommunications by Ronald S. Eward
Machine Cryptography and Modern Cryptanalysis by Cipher A. Deavours and Louis Kruh
Cryptology Yesterday, Today, and Tomorrow, Cipher A. Deavours et al., eds.
Teleconferencing Technology and Applications by Christine H. Olgren and Lorne A. Parker
The Executive Guide to Video Teleconferencing by Ronald J. Bohm and Lee B. Templeton
World-Traded Services: The Challenge for the Eighties by Raymond J. Krommenacker
The Telecommunications Deregulation Sourcebook, Stuart N. Brotman, ed.
Communications by Manus Egan
Writing and Managing Winning Technical Proposals by Tim Whalen
World Atlas of Satellites, Donald M. Jansky, ed.
Communication Satellites in the Geostationary Orbit by Donald M. Jansky and Michel C. Jeruchim
Radiodetermination Satellite Services by Martin A. Rothblatt
Communication Satellites: Power Politics in Space by Larry Martinez
New Directions in Satellite Communications: Challenges for North and South, Heather E. Hudson, ed.
Television Programming across National Boundaries: The EBU and OIRT Experience by Ernest Eugster
Evaluating Telecommunication Technology in Medicine by David Conrath, Earl Dunn, and Christopher Higgins
Optimization of Digital Transmission Systems by K. Tröndle and G. Söder
Optical Fiber Transmission Systems by Siegfried Geckeler
Traffic Flow in Switching Systems by Gerard Hebuterne
Fiber Optics for Communication and Sensors by Pradeep Wahi and K.C. Gupta
Measurements in Optical Fibers and Devices: Theory and Experiments by Giovanni Cancellieri and Umberto Ravaioli
Fiber Optics Communications, Henry F. Taylor, ed.
The Cable Television Technical Handbook by Bobby Harrell

Traffic Flow in Switching Systems

G. Hébuterne

Translated by David Oliver

Artech House
Boston and London

British Library Cataloguing in Publication Data

Hébuterne, G.
Traffic flow in switching systems.
1.Telecommunication — Switching
systems
I. Title II. Ecoulement du traffic dans
les autocommutateurs. *English*
621.3815′37 TK5103

ISBN 0-89006-231-5

Original French language edition
(Ecoulement du Trafic dans les Autocommutateurs)
published by Masson

International Standard Book Number: 0-89006-231-5
Library of Congress Catalog Card Number: 87-71224

CONTENTS

To my Parents

Preface

The purpose of the book is to survey the concepts, the problems and the tools related to the study of traffic phenomena in the switching nodes of telecommunication networks, i.e. in exchanges. Numerous benefits are to be gained from such a study: first of all, a better understanding of the dynamics of system behavior; secondly, a better design, allowing architectural choices to be based on sound concepts (and, in the end, call handling and traffic are the true criteria!); and last but not least a better dimensioning, to achieve the best grade-of-service and optimal use of equipment.

The structure and functioning of modern switching systems are described first. The emphasis is on the structures and not on technology or software, since these (although important) are of little concern here. The description is illustrated with examples drawn from current systems in use throughout the world.

The notion of *traffic* is presented, and with it all the related notions blocking, waiting, and customer behavior such as impatience or retrials.

The detailed examination of traffic phenomena — internal blocking in connecting networks, call-handling capacity and response time in common control, overload and load regulation mechanisms etc. — then begins. The main formulas and tools are presented for each of these aspects. They allow us to solve for all the dimensioning problems and to evaluate the grade-of-service experienced by subscribers.

This is "teletraffic" engineering. The word designates the study of random flows in telecommunication systems. Teletraffic is part of probability theory, and so are the methods and tools presented. A lot of work has been devoted to theoretical treatments of the subject but throughout this book we have preferred to keep a low mathematical level for clarity and simplicity.

The book cannot pretend to cover exhaustively such a vast subject as traffic in switching systems. It is hoped that it will be used as an introduction and will prompt the reader to further study. It will be of benefit to students in the telecommunications field as well as to switching system designers and the traffic specialist in charge of dimensioning exchange equipment.

The book is organized as follows. Chapters 1 and 2 present switching, its aims and constraints, and the basic traffic concepts; the main formulas and tables are brought together in the Appendix. The main body of the text is divided into two parts, the first devoted to switching networks and the second to common control. Each part presents first the general organization and functions (Chapters 3 and 7), then the problems of traffic and performance evaluation (Chapters 4, 8 and 9) and finally a deeper treatment of more difficult topics (Chapters 5 and 10); a full study of Chapters 5 and 10 may be omitted in a first reading. Problems of present research are described in Chapters 6 and 11.

Each section is illustrated as far as possible with examples of contemporary systems such as ESS4, E10, EWSD etc.

References are of prime importance for an introductory text. It would probably have been possible to double the size of the book with references. We have preferred to limit the list mainly to those papers that complement the text: full mathematical proofs, comparison with simulations, system descriptions, more intricate evaluation methods etc. can be found in the quoted references. Most of the references are to English texts but it was not possible to avoid French references completely.

The author is indebted to all his colleagues who helped him in writing the book: P. Boyer, G. Fiche, A. Gravey, J. Labetoulle, L. Romoeuf and in particular J. Roberts who clarified most of the material in Chapter 10.

G. Hébuterne

Chapter 1

INTRODUCTION

The purpose of this brief introduction is to define the position of the exchange in the general telecommunications network and, in the light of its function, to present its structure in terms of the switching network and the control units. The entire text is dedicated to the study of traffic flow problems, and the evaluation of the grade of service associated with the traffic is based on this distinction which this chapter attempts to explain and to justify as best as it possibly can.

1.1 Position of exchanges in the network

It could be said that switching was born when a third person began to use the telephone since, from then on, it was necessary to direct calls in multiple directions and it was no longer a question of connecting subscribers in twos. In fact, the modern exchange, which we are dealing with here, is simply a distant descendant of the mythical ancestor: first the manual exchange, then the rotary exchange, then the crossbar exchange and finally the electronic stored program control (SPC) exchange.

In its many forms, it essentially continues to perform the same functions: to concentrate the subscriber line traffic, to direct the calls and to structure the network. Other types of network, e.g. data communication networks, have evolved to similar structures.

Telephone users have the following characteristics.

- Their disparity. A low volume of calls is made or received by the so-called "residential" subscribers who dominate the network in terms of numbers and use their lines less than 5% of the time; the network is used more intensely by the professional subscriber who communicates between 10% and 30% of the time.
- The diversity of the services requested. Each user must be able to call any other number on the national or international network.
- The autonomy and simplicity of their terminal installation (independent of the national electricity supply) so as to minimize the problems of availability and reliability and therefore to reduce operating costs.

These three characteristics have shaped the structure of the network and the exchanges. The primary function of the latter is to serve as intermediaries for the subscribers in terms of reception of the request, its processing and establishing communication or switching to any exchange. In addition, they serve as a base for maintenance and management operations and provide "gateways" to new services.

The subscriber exchange thus defined is the basic node of the telephone network. The users are connected to it by point-to-point links; it has the responsibility of concentrating the traffic in order to optimize the use of both its equipment and the circuits connecting the exchanges.

The subscriber is inseparable from his exchange and, in order to reach him, it is first necessary to reach his exchange. The exchanges which are currently being installed can cope with between 2000 and 50 000 subscribers.

The same causes produce the same effects and there is no question of connecting all the subscriber exchanges of the network to each other, and this introduces a "hierarchy" to the network. Certain graph nodes specialize in the function of switching between subscriber exchanges; these are the transit centers which do not receive subscribers.

This process can be iterated by creating a second and then a third transit level.

National networks commonly use two transit levels (this is the case, for instance, with both the British and the French networks). The US network of the Bell System has four levels, because of its size. Transit exchanges are designated by such names as regional, sectional, primary and toll centers (in descending hierarchical order), or as secondary and primary centers, or even as class 1, class 2 etc.

Those exchanges connecting subscribers are "end offices", or "local" or "subscriber centers". Most organizations make use of "satellites" which concentrate traffic from distant clusters of customers and which are attached to a dedicated subscriber center (i.e. are unable to perform alternate routing). From our traffic point of view, private automatic branch exchanges (PABXs) are of this kind.

Communications between subscribers connected to different exchanges are either made using the upper-level transit center or by a direct route to the destination. The existence of this direct (high-usage) route depends on the volume of the traffic and results from a compromise between cost criteria, traffic criteria and security criteria. The actual telephone network is therefore a graph with a branch structure provided with a large number of direct routes (or transverse routes as distinct from the hierarchical routes) as shown in Fig. 1.1.

The final structure of the network therefore depends on the traffic, technological constraints and economic criteria.

The design, optimization and planning of the network and its management constitute a complex and multifaceted task of enormous proportions which largely lies outside the scope of the book. The interested reader should consult

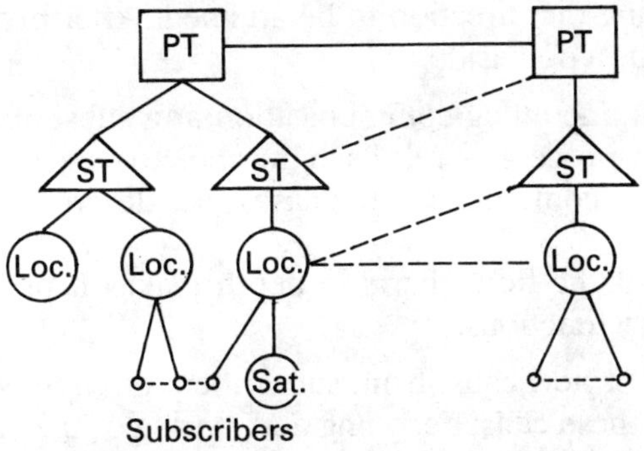

Fig. 1.1 Hierarchical structure of the network: PT, ST, primary–secondary transit centers; Loc., local office; Sat., satellite center; ——, final routes; – – –, high-usage routes (their existence is optional).

the references. We are only interested in the nodes of the graph and in particular in their "traffic flow performances".

1.2 Exchange functions

The network uses two types of exchanges: the subscriber exchange and the transit center. They both use the same components.

The general operation of the exchange differs slightly according to the type.

Type 1: subscriber center

- The exchange is notified of the request by detecting a signal (change in the state of the line). It prepares itself to receive the dialing; this is preselection.
- After receiving the request, the exchange initiates the routing of the call. If it is an outgoing call (with the destination of another center), the selection (determination and then search for the requested direction) is terminated by sending the signaling, i.e. the transmission of service information to the destination of the exchange called. The transmission is made on the same circuits as the voice circuits or on associated fixed circuits or even on specialized circuits forming a separate network called the common channel signaling network which is a real data transmission network parallel to the telephone network (the impact of this network on the exchanges will be discussed in Chapter 11).
- After the positive response of the remote center, the call is established and the conversation commences.

Type 2: transit center

In this case the preselection is initiated by a signal transmitted by the caller center and starts the incoming signaling sequence, i.e. reception of the service information including the direction to be attained. After processing, the center retransmits the request downside.

The case of an incoming communication at a subscriber center is similar except that signaling processing results in the choice of the called subscriber.

The successive communication phases are described in more detail in Chapters 3 and 7.

Examination of the flow charts in depth makes it possible to define the following elementary functions.

(a) Detection of new calls (from subscribers or circuits depending on the case); recording of these calls; decoding and storing of the called number. This function requires a device continually monitoring test points and storing interesting events. The speed of detection partly governs the overall response time and possibly also the proper functioning of the entire system, (it is necessary to detect all the events).

(b) Concentration of traffic in order to improve the efficiency of the equipment; distribution of the various flows in order to obtain a balanced effect (the traffic peaks are smoothed out by mixing, at least if the various flows are independent) and to provide a total interconnection.

(c) Processing of the calls, i.e. switching, establishment and then breaking off the communication. Each communication will result in several elementary calls of this type separated by the reaction times of the users.

(d) Charging, depending on the different tariffs in force.

(e) Surveillance of subscriber lines and circuits for monitoring proper operation; also internal surveillance; in the event of a fault, isolation of the cause and diagnostic aids.

(f) Technical management of the exchange, i.e. updating of subscriber equipment, of routing tables etc.

(g) Various services offered to the subscribers such as temporary returns and abbreviated dialing.

(h) The technical management function of primary importance, i.e. traffic measurement, which makes it possible to verify or refine predictions, to oversee the way in which the demand increases and to ensure management of the equipment. Traffic measurements also serve as the basis for planning operations.

The above-described functions are not all strictly necessary; the last three are the facilities offered by modern systems. The basic service is provided by functions (a)–(d). In fact, the history of exchanges shows a very clear trend towards the isolation of basic functions, their specification and the specialization of the corresponding equipment but also towards an increase in functions of types (e)–(g). The present computer software mainly consists of these functions, telephone tasks representing only a small proportion of programs.

All these functions can be classified under a few headings.

Functions F1. Elementary but complex functions, e.g. connection of circuits to each other, concentration and distribution. The complexity derives from the large number of sources and destinations and therefore of possible configurations; it is by nature combinatorial.

Functions F2. Advanced functions demanding high "intelligence", e.g. analysis of dialing, billing, search for routes, and fault detection. These functions handle a large number of data which they have to store in a more or less temporary fashion and they have to take decisions.

Functions F3. Intermediate functions are routine functions acting on a large number of points, e.g. monitoring subscriber lines in order to detect simple events, controlling relays etc.

1.3 Structure of the exchange

The current structure of exchanges is partly a result of tradition, i.e. the need to develop systems compatible with those of previous generations, specialization of development teams, environmental constraints etc. The classification of the different functions outlined above also justifies the current structure of exchanges.

Let us consider a "switching network" responsible for elementary tasks F1; it has been said that these tasks require a very low intelligence. They can be divided into concentration and distribution tasks. Concentration aims to increase the equipment efficiency, N sources sharing n outputs where $n < N$. The service will be satisfactory as long as the probability of total occupation remains negligible. The current trend is to entrust concentration to special units, concentrators or line switching units. Distribution is entrusted to a group selection unit. It has the function of connecting any input–output pair.

Chapter 3 describes in detail the logical structure of the network. For the moment, we shall limit ourselves to considering it as a subsystem which is capable of connecting the traffic sources to each other; depending on the technology in use, each source will be implemented by a separate electrical connection (one pair of wires or even two if the two transmission directions are separate), or several sources will share one circuit using the "time-division multiplexing" technique, which allows n channels to be carried on the same cable by sampling each channel signal and multiplexing them; moreover, each sample is binary coded in a byte. The technique is called "pulse code modulation" (PCM). Each channel is given a rate of 64 kbits/s, i.e. one time slot every 125 μs (Fig. 1.2). In the Japanese and US networks, $n = 24$ (total rate of 1.5 Mbits/s) while, in Europe, $n = 32$ of which two are reserved for synchronization and signaling purposes (total rate of 2 Mbits/s). In contrast with analog transmission, the line is unidirectional; a pair of wires has to be provided for each transmission direction.

A computer (or possibly multiprocessors) is responsible for F2 operations; the decisions to be taken are of the type "select a circuit in a given direction",

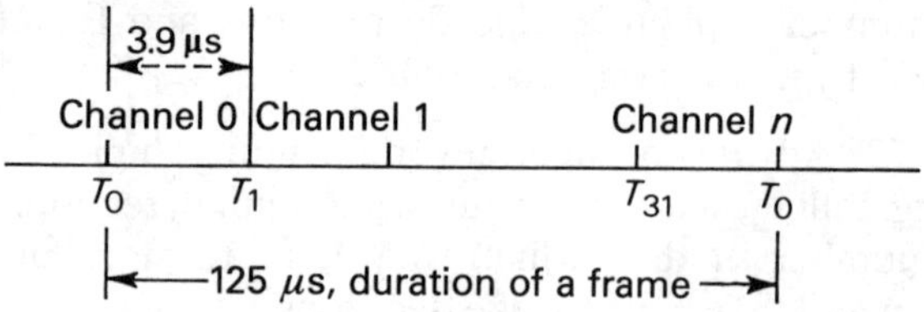

Fig. 1.2 A PCM frame ($n = 24$ or 32).

"test the condition of equipment (idle or busy)" etc. A computer is well suited to this type of activity, especially if the set of instructions about its application language is enhanced with highly specialized commands operating on binary elements. Chapter 7 describes in detail the nature of the tasks and the organization of the computer.

We have used the term "computer"; in fact, this could be a classical computer, a monoprocessor or a multiprocessor similar to those used in data-processing systems, or it could equally be a specialized structure, possibly based on processors designed for this single application. The generic term computer will be retained to cover all cases.

The so-called peripheral units responsible for F3 tasks provide the link between the computer and the external world; they may call on processors. They play a similar role to the peripherals of a data-processing system but they are adapted to the telephone environment, i.e. they can carry out surveillance of tests points and transmission of signals. Chapter 7 will explain their function.

Finally, the type of exchange that we shall study in this book has the structure shown in Fig. 1.3.

This is only a schematic diagram, the details of which will be elaborated on in the rest of this book. The distinction between the switching network and the control system should be noted. The actual position of the various peripheral units and their connection depends on the design choices and varies according to the systems; the general diagrams of such systems as METACONTA E10 [Grinsec 83] should be compared with this. Furthermore, the control system assumes various forms, e.g. a traditional centralized control computer or a distributed multiprocessor structure for the switches still to be developed; the most modern trend is to break down the control in order to bring it closer to switching and to improve performances (traffic flow and reliability).

The difference between the tasks of switching and control which results in specialization is expressed at all stages of design and development; switching and control are developed separately and many systems provide one control system linked to networks with different technologies (space-division and time-division versions of the AXE system, for example).

The evaluation of the performance, an integral part of the development of a system, also reflects this separation into two; therefore, this book has been divided into two parts in which rather different techniques are used. The next

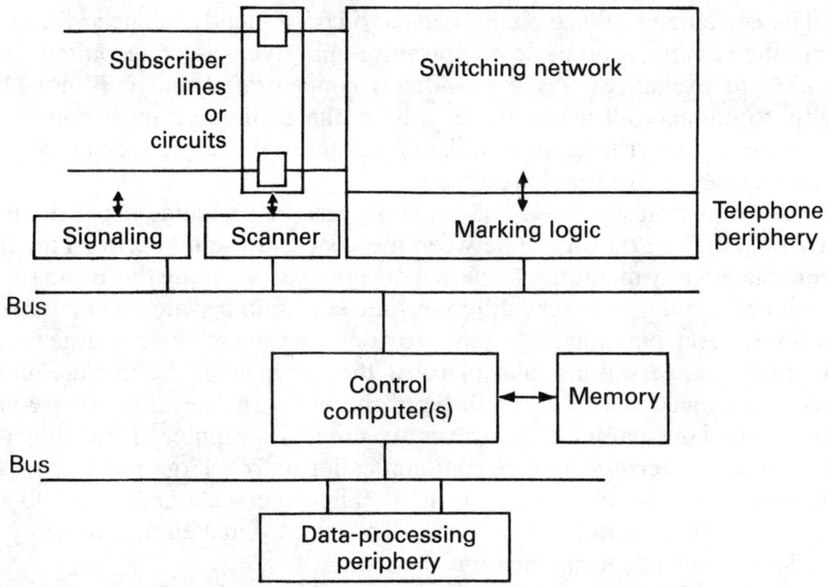

Fig. 1.3 An SPC exchange.

section will deal with the reasons for these differences and their manifestations (the clear-cut separation which is presently observed may well become less defined with technological development; however, it will probably not disappear).

1.4 Traffic flow and performance evaluation

We have discussed evaluation; this term is too vague and now should be defined.

An exchange is designed in order to provide a certain service to the user at a certain quality level. The grade of service and user-friendliness will depend on quite different factors: availability of the equipment, ergonomic factors (e.g. acoustic qualities) etc. Here we are interested in that part of the quality which is associated with the traffic, i.e. the random aspect of the user demand. The term "grade of service" is used to designate this aspect of service; standards specify its level (see below).

It will be helpful to try to understand the phenomena associated with the traffic, namely the interactions between the behavior of the users seen by the center as random and the operation of the exchange. Understanding of the traffic phenomena will make it easier to take them into account in the design, the operation and the dimensioning of equipment. This is the aim of this book.

The existence of these phenomena and consequently the need for evaluation are the results of the basic compromise underlying any telecommunication network and exchanges. As the traffic is concentrated and it is not always possible to connect all users two at a time, the equipment must compromise between economic criteria and availability criteria (the subscriber must believe that the service is entirely at his disposal).

The evaluation is carried out in two stages: the switching network and the control system. The difference between these structures and between the traffic phenomena already mentioned makes it necessary to separate these two stages.

We are dealing with very different time scales; in a switching network, the duration of a conversation (2–4 min) provides the time scale. In a large network where 5000 conversations take place at the same time, the average interval between two successive events will be of the order of 20–50 ms (these values will be justified in Chapter 2). However, the control computer breaks down each establishment or termination of communication into a large number of short elementary tasks (from 50 to 100 tasks). The time scale is 50 or 100 times shorter (elementary times of 0.1–1 ms). It would not make sense to study such dissimilar phenomena using the same model.

The switching network traditionally functions as a loss system. This means that the connection attempt which has ended in failure by "matching loss" (cf. Chapter 3) is abandoned, the user receiving a busy tone. In contrast, the establishment of a call requires repeated interventions of the control system and the periphery; in the case of unavailability, the call waits for the unit to be released.

Several arguments justify this important distinction. Firstly, there is the difference between the typical times mentioned above; the wait for a unit is of the order of magnitude of its seizure duration and a wait in the switching network (of the order of 1 min) would probably meet with opposition from the user. In the control system, each elementary task lasts only a few milliseconds (or tens of milliseconds) and the waiting times are most often not noticed by the user. Furthermore, there are a large number of practical problems to be solved in order to operate a switching network with delays, e.g. putting pending calls into reserve, passing to review and attempting to connect at each release; in contrast, waiting tasks are easy to handle in the control system. The switching network is a complex system for which an elaborate policy will be expensive; the control system is a structurally simple system for which an elaborate policy can be implemented. We shall return to this idea of complexity of a combinatorial nature in Chapter 6.

This difference in operation methods produces a difference in the properties of the traffic and the appropriate evaluation tools.

For these reasons the traffic engineer is therefore forced to separate the study of these two systems; this has the advantage of reducing their size.

Under normal conditions, this separation will introduce an absolutely negligible error. The interaction between the network and the control system is always small. The establishment and termination of connections assume a negligible time for the control system to intervene; conversely, it is assumed in the study of the control system that the switching network is transparent (no blocking).

Under serious overloading conditions, these simplifying assumptions are no longer completely true. In fact, the study of overload as presented in Chapter 10 is more aimed at verifying proper operation than giving a precise numerical definition of the transitory phenomenon. Thus there is no problem in continuing to study the two systems separately.

1.4.1 Plan of the book

The first part of the book (Chapters 3–6) studies traffic flow in switching networks. Reexamination of their structure from the traffic point of view and an introduction of the concept of matching loss are dealt with in Chapter 3. Chapter 4 reviews the methods for calculating the probability of this event; this is the main chapter which is long and complicated.

Chapter 5 treats the problem of traffic imbalances, mainly at the input stages (the mixing of sources with different characteristics on the same equipment). Chapter 6, which ends this first part, gives a glimpse into current research fields, such as calculation of a network with delay, complexity and rearrangeability.

The second part (Chapters 7–10) treats traffic flow in the control units. Chapter 7 describes the structures, also from the traffic point of view (this does not involve a description of the technology or the organization of the software, for example). The organization (centralized or distributed) of the control system will be one of the main themes of this chapter; this is the main idea behind the development of exchanges in the future.

Chapters 8 and 9 review the methods for evaluating the quality of traffic flow in the control system; e.g. deterministic methods, simulation of the environment, and analytic or simulation queuing methods.

Behavior under overload conditions is an important topic, and Chapter 10 is devoted to this. It is necessary to study the situation and to clarify the behavior of the users; references are made to the elements of a theory of reattempts.

The two parts which constitute the main body of this book are preceded by a chapter giving a brief survey of traffic concepts and their definitions (Chapter 2). This will be useful for the novice and will help with the rest of the book. For the main formulas, graphs and calculation methods which go with this introduction, see Appendix at the end of the book.

Finally, Chapter 11 contains general conclusions about the type of exchange in the future and the teletraffic problems produced by predictable developments such as multiservice networks, interconnection of processors into giant computers (more than 15 000 processors) and development of distributed control systems.

1.5 Conclusion: need for evaluation

The techniques described in the following have two complementary objectives: an aid to the design of equipment and the evaluation of existing systems (and their dimensioning).

In order to assess the performance of existing equipment before it is installed in the network, it may seem simple to organize a series of measurements. As far as the grade of service linked to traffic is concerned, this is fraught with difficulties.

First there is the need to have the sources of traffic available (the subscribers being the sole "true" source). In fact, it is necessary to wait years before an exchange achieves its nominal capacity; the load takes a long time to build up because of the development of the equipment which occurs at the same time and the cost which would be caused by a massive changeover from an old exchange to the machine to be tested. The situation is comparable with that of reliability studies where it is necessary to verify service life specifications (tens of years) from the time that the equipment is commissioned.

We still wish to emphasize that, despite the problems, it is still a real possibility to load an exchange artificially by using call-type simulators; this enables the nominal load conditions to be reproduced sufficiently well from the time of commissioning (cf. references to Chapter 8).

Next consider the difficulty of organizing and analyzing the measurements. A statistical experiment is involved which has to be prepared with as much care as an opinion poll and therefore there is no control of the procedure. To this must be added the difficulty in designing and using the measurement equipment (a large number of points needs to be monitored in real time). The advent of computers does not entirely solve this problem since the measurement process disturbs the traffic flow.

A solution consists in simulating subscriber behavior; this is environment simulation (see Chapter 8). However, if it is possible to simulate the environment, is it not possible to simulate the machine also?

A series of measurements will be extremely expensive. The evaluation "on paper" based on mathematical models, which have been necessarily simplified, constitutes a powerful aid in the process of overall qualification and in dimensioning the units. This does not make the measurements described above redundant but it can shed light on them and provide "stepping stones" such as what should be observed and what is expected to be found.

In addition, it should be noted that the evaluation of existing equipment is only a last resort, an *a posteriori* verification of the objectives of the designers whose disappointment if the measurements invalidate the predictions will only be equal to that of the financial departments of the company.

In other words, a grasp of the traffic phenomena during design is one of the keys to success. It would be pointless and absurd to expect that the exchange is completed in order to compare the respective merits of one priority rule or another for assigning tasks to the central processor. The role of the traffic specialist in the development team must be to participate in the process of selecting between competing solutions by providing a comparative performance evaluation (commonsense, if it can help, will not be sufficient to make the right choice; a solid formula and rapid simulation will be worth more).

In this connection, simplified models turn out to be effective, rapid to implement and sufficiently precise. Lack of reliable numerical data makes an elaborate model useless. Sophisticated models will be reserved for the final evaluation.

As far as the final evaluation of existing systems is concerned, it is above all a matter of verifying the conformity of the exchange to the user's specifications.

Most countries have standards whose format usually reflects the exchange organization, especially the separation between the switching (speech) network and control system. The Consultative Committee for International Telegraph and Telephone (CCITT) has developed a standard for international calls (recommendation E543) which says that the probability of lost calls by "internal blocking" in the switching network must be less than 0.002 under nominal loading conditions, and 0.01 in the event of overload (20% increase in traffic).

The values for control systems set limits for the delays or rather for the quantiles of their distributions. For instance, the CCITT recommendation requires that the "preselection time" (the time between the subscriber unhooking and receiving the dial tone) be less than 0.5 s in 95% of the cases for nominal load, and less than 1 s (again $P(\text{delay} > 1 \text{ s}) < 0.05$) for overload conditions, defined here by an increase in 40% of the number of processed calls. Similar values specify the selection duration.

Finally, an availability constraint is usually defined. It results from reliability studies but is intimately related to traffic flow; availability criteria lead to the practice of redundancy in the hardware in order to ensure the specified quality of traffic flow under degraded conditions.

The significance of the concepts and values mentioned here will be explained later in the book. Note the difference in the definition of overload conditions which will make sense in Chapter 10 (user behavior).

The reader is referred to the complete text of the standards of his choice for the entire set of specifications and the conditions for their application.

References

There are relatively few books describing modern switching systems. In [Collins 73], [Flowers 76], [Hills 76], [Inose 81], [Pearce 81] and [Grinsec 83], various technological or software aspects not discussed in the present text are dealt with.

As far as the position of the exchange in the network is concerned, and the related very important aspects of network planning, a good summary is provided by the COST 201 studies (see for example [Dresser 83]). The French telephone network and the traffic engineering principles on which it is based are discussed in [Lacout 82].

The history of electronic switching, from its beginning until the present day, is described in a few papers by P. Lucas (in French). See [Lucas] in Chapter 7.

[Collins 73] A. Collins, R. D. Pedersen. *A Time for Innovation*, Merle Collins Foundation, Dallas, Texas, 1973.

[Dresser 83] J. Dresser, J. A. C. Gomes, R. J. Mentel, J. M. Mepuis, E. J. Sara. COST 201 a European research project. A flexible procedure for minimizing the costs of a switched network. *Proceedings of the 10th International Teletraffic Congress*, Montreal, 1983, Paper 2.1.3.

[Flowers 76] T. H. Flowers. *Introduction to Exchange Systems*, John Wiley, New York, 1976.

[Grinsec 83] Grinsec. *Electronic Switching* (2 volumes), *Studies in Telecommunications*, Vol. 2, North-Holland, Amsterdam, 1983.

[Hills 76] M. T. Hills, S. Kano. *Programming electronic switching systems*, Peter Peregrinus, London, 1976.

[Inose 81] H. Inose, *An Introduction to Digital Integrated Communications Systems*, Peter Peregrinus, London, 1981.

[Lacout 82] Lacout, Cade. *Les Télécommunications Françaises*, Ministère des PTT, 1982.

[Pearce 81] J. G. Pearce. *Telecommunication Switching*, Plenum Press, New York, 1981.

The following chapters contain additional specialized references.

Chapter 2

ELEMENTARY TELETRAFFIC CONCEPTS

Throughout this book, we shall study complex systems with a perfectly defined operation but subject to a random environment. In this chapter, we solve the following elementary problem which will apply to everything that follows.

Let us consider a group of R "servers" (trunks connecting two switches, say) handling a flow of "customers" (telephone calls). This flow may originate from a finite group of N sources or from a very large number (infinite in the case of circuits). In all cases, there is the possibility of total occupation and also of rejection or of queuing (depending on the operating model). What will be the probability of such an event?

Care should be taken not to confuse elementary with simple. The complete treatment of this problem remains to be carried out and thousands of pages have been dedicated to it. In the language of queuing theory, it is a matter of studying the $G/G/R$ or $G/G/R/R$ system (the notation will be explained later). The study of the simplest cases leads us to define the following.

- The arrival process of the customers (see Section 2.1).
- The service process (see Section 2.2).
- The service mechanisms (see Section 2.3).

We shall apply the method of "birth and death processes" (Section 2.4) in order to calculate the behavior of the simplest systems such as the case of traffic originating from a very large number of sources (Erlang traffic in Section 2.6) or from a limited number of sources (Engset traffic in Section 2.7) with lost calls cleared.

In Section 2.8, some special formulas are established which are used in Chapter 4; finally, Section 2.9 gives an introduction to queuing systems (based on the simplest case, the $M/M/R$ queue).

The terms "customers" and "calls" will be used in this chapter indistinguishably to designate the entities occupying the resources which will be called "circuits" or "servers". These are equivalent terms originating from the language of queuing theory and the more traditional language of telecommunications. Concepts specific to teletraffic* will be defined in Section 2.5.

This chapter discusses briefly a vast subject in a few pages. The treatment is obviously not exhaustive. Rather it should be regarded as a compilation of definitions and concepts which are used throughout the remainder of the book as well as a statement of the assumptions about traffic on which the entire book is based. A low-level approach has been deliberately adopted in terms of formal mathematics; we refer the reader to the bibliography for a more exhaustive and formal treatment. Nevertheless, this introduction should make it possible to read and understand the rest of this work.

2.1 The arrival process

The times of appearance of the calls can be seen in Fig. 2.1.

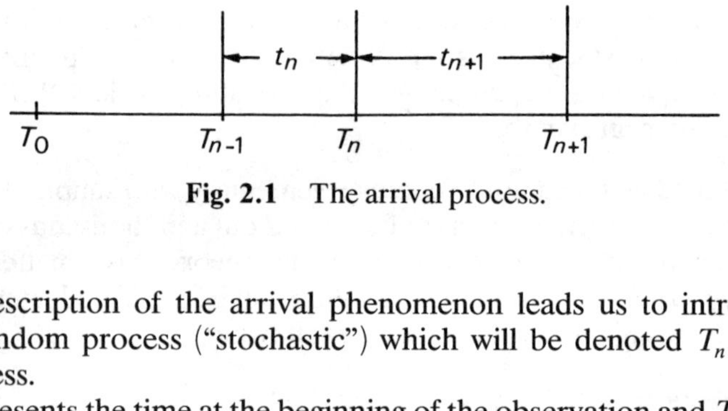

Fig. 2.1 The arrival process.

The description of the arrival phenomenon leads us to introduce, as a model, a random process ("stochastic") which will be denoted T_n $(n>0)$, the arrival process.

T_0 represents the time at the beginning of the observation and T_n $(n \geqslant 1)$ the time of arrival of the nth call. We denote by t_n the time interval between the times of appearance of the $(n-1)$th and nth calls:

$$t_n = T_n - T_{n-1}$$

In the general case, t_n will depend on the entire past of the process. For simplicity, we shall limit ourselves to the case when the variables t_n are independent and have the same distribution; such a process will be called a renewal process. Let

$$A(t) = P(t_n = T_n - T_{n-1} \leqslant t)$$

*The term "teletraffic" is taken to mean a restriction of the "traffic" to telecommunication systems (as opposed to railway or air traffic for instance).

The Poisson process is a remarkable example of renewal and plays a fundamental role in the study of queuing and traffic phenomena. It is based on the following assumptions.

Assumption A1. The probability of the appearance of a new call during a time interval $(t, t+\Delta t)$ does not depend on what happened before t.

Assumption A2. The probability of the appearance of a call during t is proportional to Δt, to the first order of magnitude, and the probability of appearance of another call is negligible.

Let us denote by $P_k(\theta)$ the probability that there are exactly k appearances in the interval θ:

$$P_1(\Delta t) = \lambda\, \Delta t + O(\Delta t) \tag{2.1}$$

$$1 - P_0(\Delta t) - P_1(\Delta t) = O(\Delta t) \tag{2.1'}$$

These two assumptions are sufficient to calculate $A(t)$. Let us first evaluate $P_k(t)$. According to the above assumptions,

$$P_{k+1}(t+\Delta t) = P_{k+1}(t)(1-\lambda\,\Delta t) + P_k(t)\lambda\,\Delta t + O(\Delta t)$$

We let Δt tend to 0; the preceding relationship implies that $P_k(t)$ possesses a derivative and the resulting equation is simply solved:

$$P_k(t) = \frac{(\lambda t)^k}{k!}\,\mathrm{e}^{-\lambda t} \tag{2.2}$$

In particular,

$$P(t_1 > t) = P_0(t) = \mathrm{e}^{-\lambda t}$$

$$A(t) = 1 - \mathrm{e}^{-\lambda t}$$

The coefficient λ is the "arrival rate" of the process; this terminology is justified by the equation relating it to the mean time between successive arrivals:

$$E(t) = \int t\,\mathrm{d}A(t) = \frac{1}{\lambda}$$

The variance of this time also only depends on λ:

$$\sigma^2 = E(t^2) - \{E(t)\}^2 = \frac{1}{\lambda^2}$$

(it should be noted that σ, the standard deviation, is equal to the mean).

The Poisson process is also known as the "memoryless process". This basic property explains its importance.

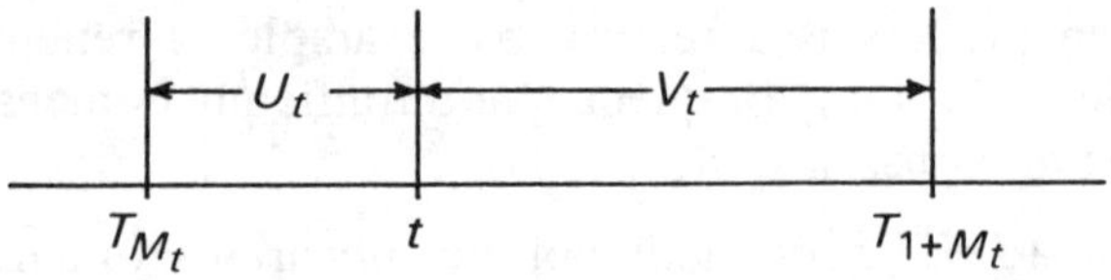

Fig. 2.2

We observe the process at time t; M_t arrivals have already occurred. We introduce the variables U_t, the time elapsed since the last arrival, and V_t, the time until the next arrival (Fig. 2.2).

Theorem 1 (Feller). The variable V_t has an exponential distribution with the parameter λ.

Intuitively, V_t has the same distribution as t_n; irrespective of the time which has already elapsed since the last arrival, the probability of still having to wait x time units remains equal to $e^{-\lambda x}$.

In particular, an observer examining the process at any time t has a probability

$$dP = \lambda\, dt + O(dt) \tag{2.3}$$

of seeing an arrival between t and $t+dt$.

In practice, streams of Poissonian arrivals are much more often observed than would be expected from the basic assumptions A1 and A2. In fact, the following theorem (the so-called Palm–Kintchine theorem) shows that an arrival process resulting from the superimposition of a very large number of any independent processes is Poissonian.

Theorem 2 (Cinlar). Let $\mathscr{P}_1, \mathscr{P}_2, \ldots, \mathscr{P}_k$ be k independent renewal processes. We denote by λ_i the arrival rate (the reciprocal of the mean time between arrivals) of the process i. Let $\mathscr{P}_n$ be the process formed by the superimposition of the $\mathscr{P}_i$ and let $\lambda^* = \Sigma_j \lambda_j$.

If $k \to \infty$ and $\lambda \to \lambda^*$, then $\mathscr{P}_n$ tends towards a Poisson process with the parameter λ^*.

For example, the arrival of communications offered to a link connecting two transit centers results from the superimposition of the offers of each subscriber having access to the link. Each user offers infinitesimal traffic which can be assumed to be independent of the other sources; a Poisson process is observed. Similarly, calls offered to the control system of an exchange originating from all the subscribers of the center have this distribution (see the references to Chapter 8 for the experimental evidence for this).

Note that in the general case the superimposition of two renewals does not give a renewal process (Fig. 2.3).

The time T_4 does not depend on T_3 but on T_2. In the general case, T_n will depend on all the preceding T_i: T_{n-1}, $T_{n-2}, \ldots$. In the case when the components

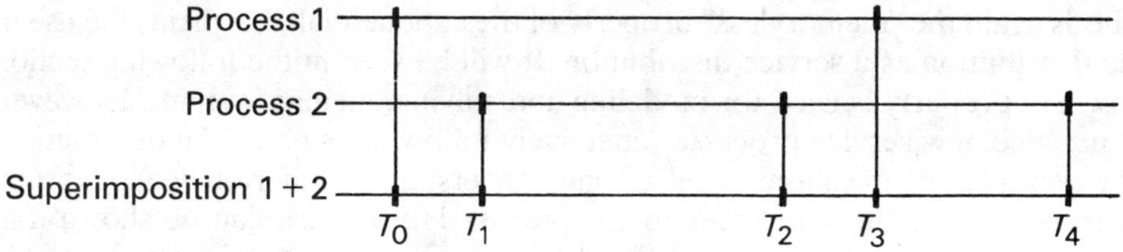

Fig. 2.3 Superimposition of renewals.

of finite number are Poissonian, and only in this case, is the resultant a renewal (and what is more, Poissonian).

2.2 Service processes

In the same manner, the most general service process will be extremely complex depending on the preceding services, the arrivals etc. As for the arrivals, we shall limit ourselves to the case when the service time may be described by a random variable obeying the distribution $F(x)$ which is the same for each customer, the service times for successive customers being chosen as independent (these are usually referred to as independent and identically distributed variables). From F the mean and the variance can be deduced:

$$E(s) = \int t \, \mathrm{d}F(t)$$

$$\operatorname{var}(s) = E(s^2) - \{E(s)\}^2 = \sigma^2$$

The coefficient of variation is the ratio $c = \sigma(s)/E(s)$.

The simplest distribution to describe is the constant when $c = 0$. However, the exponential distribution, characterized by

$$F(x) = 1 - \mathrm{e}^{-\mu x}$$

with mean

$$E(s) = \frac{1}{\mu} \qquad \sigma^2 = \frac{1}{\mu^2} \qquad c = 1$$

is frequently used.

If a service lasts for t_0s, the probability that it terminates in the next second does not depend on t_0. Thus the probability that the service ends between t and $t + \mathrm{d}t$ is simply equal to

$$\mathrm{d}P = \mu \, \mathrm{d}t + O(\mathrm{d}t) \tag{2.4}$$

This is again the "memoryless" property of the exponential; it explains the use of the distribution as a service distribution. It will be seen in the following section how this property is used for modeling and solving queuing systems. However, in practice, few service processes effectively follow an exponential distribution. For example, the durations of telephone conversations rather obey a lognormal distribution (see the references to Chapter 8). However, it can be shown that some results are independent of the shape of the service distribution and only depend on its mean.

Nevertheless, for waiting systems, the service distribution plays an important part and it is advisable to take care in using the exponential distribution (e.g. the Pollaczec formula); a precise numerical result will not be expected without the exact service distribution (cf. Section 2.9).

2.3 Service mechanism; the Little formula

We have described the arrival process of customers by a distribution $A(t)$ and the service process by a distribution $F(t)$. The system will be entirely specified if R, the number of servers, and their operating mode are defined. Most devices obey the following rules.

- The servers begin to operate as soon as a customer is present (waiting customers and an idle server cannot exist at the same time).
- The customers are served in the order of their arrival (this is called first in, first out (FIFO)). There are other more complex procedures.
- When the R servers are active, the incoming customers cannot be served immediately. A distinction is made between waiting systems where these customers are allowed to wait in front of the servers and loss systems if there is no waiting room; in this case, excess clients are lost and disappear finally. The terminology "lost calls cleared" is often used since customers not admitted definitely disappear; for simplicity, however, we shall use the term "loss systems". Of course, any intermediate configuration can be imagined (R servers and $N > R$ places, i.e. $N-R$ queuing positions, for queuing with limited capacity). In telephony, calls never disappear totally; in this case, we talk about reattempts (see Chapter 10).

Kendall's notation summarizes in symbolic form the principal characteristics of the system defined in this way:

$$A/B/m/K$$

where A designates the arrival process, B designates the service distribution (M denotes the exponential (memoryless or Markovian) distribution, D the constant-duration (deterministic) distribution, E_k the Erlang k distribution which is a distribution of time for traversing k exponential servers with the same parameter in cascade, G the most general distribution and GI the most general distribution for which the durations are independent and identically distributed), m designates the number of servers and K designates the total number of waiting plus service places (if omitted, K is assumed to be infinite).

Thus the queue $GI/M/1$ designates a system with independent and identically distributed arrivals (renewal), with exponential service and a single server. $M/M/R/R$ designates a queue with Poisson arrivals, and an exponential service operated as a loss system.

The following sections should explain some of the mystery surrounding the $M/M/R$ systems. For the most general $G/G/R$ system, moreover, we shall give a very important result, the Little formula.

Let L be the average length of the queue (the number of customers waiting), W be the average waiting time and λ the arrival rate (the reciprocal of the average time between successive arrivals). Then,

$$L = \lambda W$$

which is the formula due to Little. This formula remains valid for other interpretations of the variables. It is in fact a conservation law. Thus, L and W can refer indistinguishably to waiting customers or to all the customers (both waiting and in service) in the queue. Similarly, the "system" may be a waiting system or a loss system (see Section 2.5, for example).

2.4 Birth and death process

In order to describe a service system (say the $M/M/1$ queue) it is necessary to know the following.

- The number of customers being serviced or waiting.
- The remaining service time to be used by the customer being serviced.
- The time up to the next arrival.

In fact, the property of the exponential distribution (memoryless) makes it possible to simplify this description by reducing the number of customers in the system. This number alone specifies the state of the queue, i.e. it allows its evolution to be predicted, in terms of probability. Thus, if n clients are present at time $t\,(n>0)$, then

$$\begin{array}{ll} \text{for a probability } \lambda\,\mathrm{d}t, & \text{there will be } n+1 \text{ at } t+\mathrm{d}t \\ \text{for a probability } \mu\,\mathrm{d}t, & \text{there will be } n-1 \text{ at } t+\mathrm{d}t \\ \text{for a probability } 1-(\lambda+\mu)\,\mathrm{d}t, & \text{there will be } n \text{ at } t+\mathrm{d}t \end{array} \qquad (2.5)$$

This follows obviously from eqns (2.3) and (2.4).

It is convenient to represent this model graphically (Fig. 2.4) by making the possible transitions explicit. Each circle represents a state ($n=0$ indicates system empty and server at rest; $n=1$ indicates one client in service; ...); an arrow joining $n=k$ to $n=k+1$ symbolizes the arrival of a client who increases

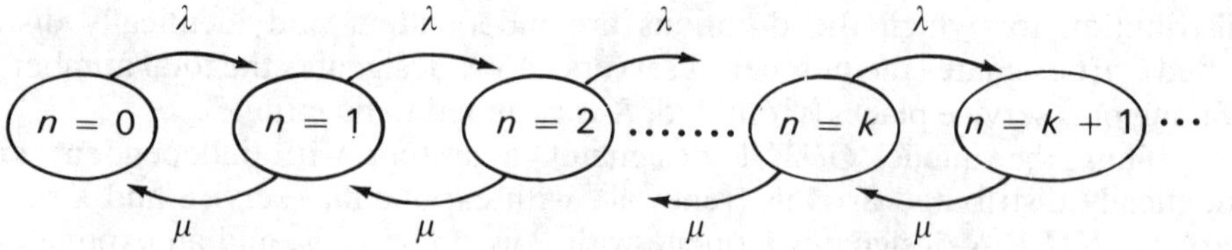

Fig. 2.4 A birth and death process (the $M/M/1$ queue).

the number of clients from n to $n+1$; the occurrence rate of the event is indicated by λ, and the occurrence rate of the service by μ.

The state equations describe the evolution of the process. Let $P_k(t)$ be the probability of finding the system in the state k at time t. We write

$$\begin{aligned}P_k(t+\mathrm{d}t) &= P_k(t) + \frac{\mathrm{d}P_k}{\mathrm{d}t}\,\mathrm{d}t \\ &= P_k(t)(1-\lambda\,\mathrm{d}t-\mu\,\mathrm{d}t) \\ &\quad + P_{k-1}(t)\lambda\,\mathrm{d}t + P_{k+1}(t)\mu\,\mathrm{d}t \qquad k>1 \\ P_0(t+\mathrm{d}t) &= P_0(t)(1-\lambda\,\mathrm{d}t) + P_1(t)\mu\,\mathrm{d}t\end{aligned}$$

which can be immediately understood; the system is in the state k at $t+\mathrm{d}t$ if it was there at t and if nothing has happened, or if it was in the state $k-1$ and an arrival occurred, or finally if it was in the state $k+1$ and a departure occurred. The reader should note the use of eqn (2.5) from which these relationships derive.

We shall limit ourselves to the study of the steady-state regime, assuming for the time being that it exists (see eqn (2.8)); the probabilities no longer depend on time ($\mathrm{d}P_k/\mathrm{d}t=0$) and the following can be derived:

$$\lambda P_0 = \mu P_1$$

$$\vdots$$

$$(\lambda+\mu)P_k = \lambda P_{k-1} + \mu P_{k+1} \qquad k>1$$

$$\sum_{k\geqslant 0} P_k = 1$$

The last equation, called the normalization equation, specifies that the system must always be in a certain state.

We note that the equations can be derived from the diagram in Fig. 2.4; the probability flux entering a state $\lambda P_{k-1}+\mu P_{k+1}$ must be equal to the flux which leaves it, which is $(\lambda+\mu)P_k$, here in order to guarantee a steady-state condition where the P_k remain constant.

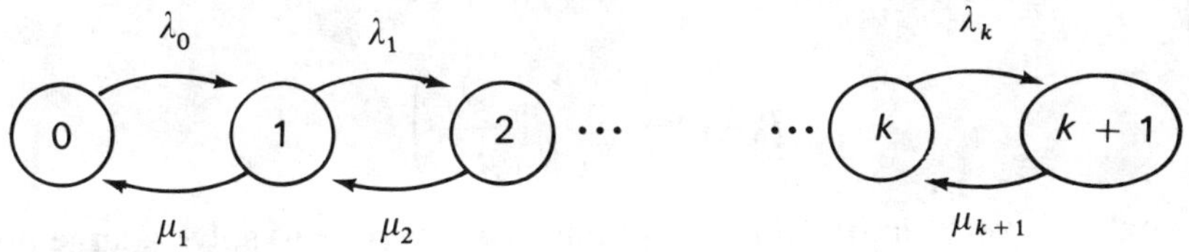

Fig. 2.5 The general birth and death process.

The recursion solution gives

$$P_k = \left(\frac{\lambda}{\mu}\right)^k P_0$$

$$P_0 = \left\{1 + \frac{\lambda}{\mu} + \ldots + \left(\frac{\lambda}{\mu}\right)^k + \ldots\right\}^{-1} = 1 - \frac{\lambda}{\mu}$$

λ/μ which is traditionally designated ρ is interpreted as the utilization rate, i.e. the probability that the server is occupied or rather the proportion of time during which it is active. This will be the "offered traffic" (see Section 2.5).

The method can easily be extended to the most general case of birth and death processes where the coefficients λ and μ depend on the state (Fig. 2.5).

As above, this diagram is useful for writing the state equations

$$\begin{aligned} &\lambda_0 P_0 = \mu_1 P_1 \\ &\vdots \\ &(\lambda_k + \mu_k) P_k = \lambda_{k-1} P_{k-1} + \mu_{k+1} P_{k+1} \end{aligned} \tag{2.6}$$

and also the normalization equation

$$\sum_{k \geqslant 0} P_k = 1 \tag{2.6b}$$

The solution of this system (2.6), provided that it exists, is given by (2.7):

$$\begin{aligned} &P_1 = \frac{\lambda_0}{\mu_1} P_0 \\ &P_2 = \frac{\lambda_0}{\mu_1} \frac{\lambda_1}{\mu_2} P_0 \\ &\vdots \\ &P_k = \frac{\lambda_0}{\mu_1} \frac{\lambda_1}{\mu_2} \cdots \frac{\lambda_{k-1}}{\mu_k} P_0 = \left(\prod_{i=0}^{k-1} \frac{\lambda_i}{\mu_{i+1}}\right) P_0 \end{aligned} \tag{2.7}$$

with

$$P_0 = \left(1 + \sum_{k \geqslant 1} \prod_{i=0}^{k-1} \frac{\lambda_i}{\mu_{i+1}}\right)^{-1}$$

Although it is very simple, this model makes it possible to solve a large number of interesting problems by a suitable choice of the coefficients λ_i and μ_i.

This system of equations always allows a unique solution provided that it is possible to verify the normalization condition (2.6b). This is so in particular if the number of states is finite or if there exists a value k_0 such that (sufficient condition)

$$k > k_0 \rightarrow \frac{\lambda_k}{\mu_{k+1}} \leqslant \alpha < 1 \tag{2.8}$$

In this case, P_k will be interpreted as the proportion of the time which the system spends in the state k.

2.5 Definition of teletraffic terms; traffic and blocking

Within the framework of our elementary problem we shall define the main concepts of teletraffic used in the following; their general usage is beyond the scope of the book. A more formal mathematical approach is possible (cf. [Syski 86], Chapter 4). When the objectives of the book and the space available are borne in mind, it seemed sensible to adopt a more intuitive approach even if it entailed referring the reader to the references for a formal presentation and for complete proofs.

The following variables describe the behavior of the sources: T, the average service time (for the exponential distribution, $T = 1/\mu$); C, the arrival rate, which is the average number of calls appearing in unit time or the reciprocal of the average time between successive arrivals; C_0, the effective arrival rate, which is the average number of calls accepted.

For a system governed by a birth and death process, the state probabilities will be denoted by P_j, $j = 0, 1, \ldots, R$ (R may be infinite).

Telephonists make a careful distinction between the following concepts.

Offered traffic. This is the average number of calls arriving during the average service time. It is designated A. With the above definitions,

$$A = CT \tag{2.9}$$

Carried traffic. This is the average number of servers (circuits) occupied at the same time. It is designated A_0 [Syski 86]. If the state probabilities have been defined, it is written

$$A_0 = \sum_{j \geqslant 0} jP_j$$

The Little formula makes it possible also to write

$$A_0 = C_0 T \tag{2.10}$$

The carried traffic can also be interpreted as the average number of customers accepted during the duration of service. It is measured by summing the durations of occupation of the servers and dividing by the total observation time.

The distinction between offered traffic and carried traffic can be easily interpreted; in a system without loss, all arrivals are accepted, i.e. $C_0 = C$ and $A_0 = A$. In other words, in a time θ, $C = A\theta/T$ clients arrive who occupy the servers for a time $A\theta$, i.e. an occupation rate of A. In this case, $A = A_0$ (this is the ρ defined in Section 2.4). In other cases, $A_0 < A$. It will be seen in the examples given in the following sections how C and C_0 are evaluated in practice.

Following the CCITT recommendations, traffic is measured in erlangs, from the name of the Danish engineer and mathematician who pioneered teletraffic (1878–1929). The erlang is the traffic carried by a line which is permanently busy (or by a group of n lines totaling 60 min of cumulative occupation per hour). Similarly, a server seeing one communication arriving on average per period aT will be subjected to a carried traffic of $1/a$ erlangs.

The "traffic unit" (abbreviated to TU) is another name for the "erlang". The "hundred call seconds per hour" (abbreviated to CCS) is sometimes used to measure traffic, C being expressed in calls per hour and T as a multiple of 100 s; $A = CT$ expresses the traffic in CCS, and thus 1 erlang = 36 CCS.

Time congestion. This is the proportion of time (designated B_T) during which all the servers are busy. For a birth and death process, $B_T = P(R)$, the total occupation probability.

Call congestion. Conversely, the call congestion B designates the proportion of rejected calls (in a "lost call cleared" system):

$$B = (C - C_0)/C = (A - A_0)/A \tag{2.11}$$

The distinction between B and B_T is important. The users experience B but measurement methods often supply B_T. In the particular case of Poisson arrivals, B_T and B are equal but this is not true, in general.

These definitions apply to all types of arrival or service distribution. We shall study two particular cases which are very useful in the following.

2.6 Erlang traffic

This incorrect term here designates a system with limited capacity (R servers and R places) functioning with "lost calls cleared" and subjected to an arrival flow obeying the Poisson distribution. Exponential occupation times are assumed which makes it possible to utilize birth and death processes. In reality, the results obtained are valid for the $M/G/R/R$ queue.

The state of the system is characterized by n, the number of active servers ($0 \leqslant n \leqslant R$). It is necessary to find P_n, the state probabilities of the system. This first example makes it possible to illustrate the difference between λ, C and C_0. The rate of the associated birth process depends on the state:

$$\lambda_i = \lambda \qquad \text{for } 0 \leqslant i < R$$
$$\lambda_R = 0$$

Because of the limited capacity, the calls arriving during the stay in state R are lost. Nevertheless, according to the definition,

$$C = \lambda$$

whereas C_0 can be written as

$$C_0 = \Sigma \lambda_i P_i = \lambda(1 - P_R)$$

The offered traffic is

$$A = CT$$

and

$$A_0 = A(1 - P_R)$$

The time congestion B_T is equal to P_R and, according to eqn (2.11),

$$B = \frac{A - A_0}{A} = P_R$$

As mentioned above, the Poisson character of the arrivals implies that B and B_T are equal.

The solution of the birth and death equations gives us P_n and, in particular, P_R. Here, the rates μ_k are equal to $\mu_k = k\mu$.

If indeed k servers are active at t (these are conditional events with this assumption),

$$\begin{aligned} &P(\text{a departure between } t \text{ and } t + \mathrm{d}t \mid k \text{ servers active at } t) \\ &\quad = P(\text{server 1 terminates}) + P(\text{server 2}) + \ldots + P(\text{server } k \text{ terminates}) \\ &\quad = k\mu \, \mathrm{d}t \end{aligned}$$

This is because the servers are independent of each other and of the exponential property (infinitesimals of higher order have been neglected). The diagram in Fig. 2.6 will help to write the equations of state.

We note that this system of $R + 1$ equations with $R + 1$ unknowns is of rank R and cannot be solved by itself (this can be seen by summing the terms on the left-hand side and on the right-hand side); the normalization equation is necessary in order to complete the solution. This is

$$P_j = \frac{A^j/j!}{1 + A + A^2/2! + \ldots + A^R/R!} \tag{2.12}$$

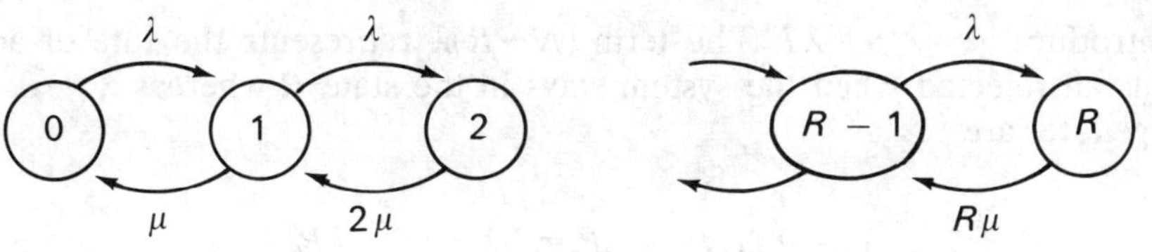

Fig. 2.6

$$B = B_T = P_R = \frac{A^R/R!}{1 + A + A^2/2! + \ldots + A^R/R!} = E(R, A) \tag{2.13}$$

This is the famous Erlang formula, often referred to as the Erlang B formula. It is calculated by recursion and is also given in the form of graphs or charts (see Appendix at the end of the book). As mentioned above, eqns (2.12) and (2.13) remain valid for a general service process but the proof is not as obvious.

When R becomes very large (A/R is very small), we can write

$$P_j = \frac{A^j}{j!} e^{-A} \tag{2.14}$$

and of course $P_R = 0$. Such a distribution is sometimes called Poisson traffic.

Note that the term "Erlang traffic" has been used. This designates the process of Poisson arrivals with a constant rate; as will be seen in the next section, this is the traffic offered by an infinite number of users. Sometimes in the German and English literature, the term "pure chance traffic of type 1" (abbreviated to PCT 1) is found.

2.7 Engset and Bernoulli traffic

Here we shall assume a system with limited capacity (R servers and R positions) operated with lost calls cleared and carrying the traffic from a limited number N of sources (for a typical application of a subscriber stage, see Chapters 4 and 5).

The behavior of a source must be described exactly. Obviously, a source which is already occupied does not transmit any new calls. The rate of appearance of calls coming from an idle source is therefore defined as λ (according to eqn (2.3), $\lambda\, dt$ is the probability that an inactive source at t will transmit a call between t and $t + dt$). Since busy sources no longer generate new calls, the birth and death coefficients of the system in eqn (2.6) are

$$\lambda_k = \lambda(N - k) \qquad k = 0, 1, \ldots, R-1$$

$$\lambda_R = 0$$

$$\mu_k = k\mu = \frac{k}{T}$$

We introduce $\alpha = \lambda/\mu = \lambda T$. The term $(N-R)\lambda$ represents the rate of actual arrivals all rejected when the system stays in the state R whereas $\lambda_R = 0$. The average rates are

$$C = \sum_{k=0}^{R} (N-k)\lambda P_k = N\lambda - \lambda \sum_{k=0}^{R} kP_k$$

$$C_0 = \sum_{k=0}^{R-1} (N-k)\lambda P_k = C - (N-R)\lambda P_R$$

where P_j, $j = 0, 1, \ldots, R$ designates the associated steady-state probability. The offered traffic will be defined in accordance with eqn (2.9) by

$$A = CT = \alpha\left(N - \sum_{j \leqslant R} jP_j\right) = \alpha(N - A_0) \tag{2.15}$$

and the carried traffic in accordance with eqn (2.10) by

$$A_0 = \sum_{j=0}^{R} jP_j = A - (N-R)\alpha P_R$$

The call congestion and the time congestion differ:

$$B_T = P_R$$

$$B = (C - C_0)/C = (N-R)\lambda P_R/C \tag{2.16}$$

$$= \frac{N-R}{N-A_0} P_R$$

The equations of state are easily solved; their solution is*

$$P_j = \frac{\binom{N}{j}\alpha^j}{\sum_{k=0}^{R}\binom{N}{k}\alpha^k} \tag{2.17}$$

i.e. in particular from eqn (2.16) after a few manipulations,

---

*We denote $\binom{j}{k}$ as the binomial coefficient. This is the number of ways of choosing k objects from j:

$$\binom{j}{k} = \frac{j!}{k!(j-k)!}$$

$$B = \frac{\binom{N-1}{R}\alpha^R}{\sum_{k \leqslant R}\binom{N-1}{k}\alpha^k} \tag{2.18}$$

It should be noted that λ (and therefore α) cannot be measured directly. It has been defined as the transmission rate for an idle source. Only the average transmission rate c of a source can be observed; we have $c = C/N$.

We set $a = c/\mu = A/N$ (offered traffic per source); $1/c$ represents the average interval between the commencement of communication attempts (Fig. 2.7).

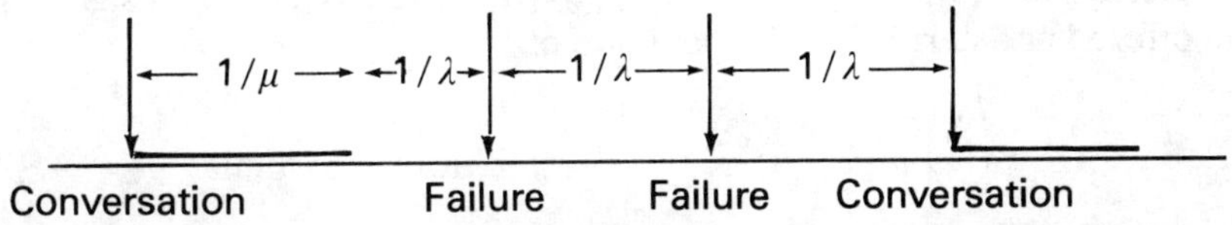

Fig. 2.7

Depending on the fate of the attempt (failure with probability B; otherwise success), the interval between successive attempts will be increased by the duration of conversation or not, i.e. on average $1/\mu$. The average interval $1/c$ is therefore given by

$$1/c = \left(\frac{1}{\mu} + \frac{1}{\lambda}\right)(1 - B) + \frac{1}{\lambda} B$$

from which

$$c = \frac{\lambda}{1 + \alpha(1 - B)}$$

$$a = \frac{A}{N} = \frac{\alpha}{1 + \alpha(1 - B)} \tag{2.19}$$

It is usually assumed that either the average traffic offered by source c or the total traffic A is given. Equations (2.17), (2.18) and (2.19) make it possible to solve the problem in question indirectly (it is necessary to eliminate α).

In many applications, low congestion is expected. We then write

$$a = \frac{\alpha}{1 + \alpha} \qquad \alpha = \frac{a}{1 - a} \tag{2.20}$$

expressions which can be substituted in eqn (2.17).

There are tables which give the values for congestion in eqn (2.17); it can also be evaluated by recursion (see Appendix at the end of the book). As in the Erlang case, these formulas are still valid for any service distribution. We speak of the Engset formulas or Engset traffic (in the same sense as in the preceding section) and equally of "pure chance traffic of type 2" (abbreviated to PCT 2).

An important particular case is $R \geqslant N$. There is obviously then no rejection. We have

$$P_j = \frac{\binom{N}{j}\alpha^j}{\sum_{k \leqslant N}\binom{N}{k}\alpha^k} = \binom{N}{j}\frac{\alpha^j}{(1+\alpha)^N}$$

This is Bernoulli traffic. It uses $a = \alpha/(1+\alpha)$ since $B = 0$; a designates the average offered and carried traffic per source:

$$P_j = \binom{N}{j} a^j(1-a)^{N-j} \qquad \text{(binomial formula)} \qquad (2.21)$$

which clearly shows that the servers can be considered as independent (in addition, the independence makes it possible to obtain eqn (2.21) directly). Equation (2.21) is of vital importance; we can often approximate eqn (2.17) by eqn (2.21), in particular when N/R is not too large.

2.8 Some useful formulas

As an example, let us take a look at Erlang traffic ($M/M/R/R$ queue). Equation (2.12) gives, under steady-state conditions, the probability of finding j servers busy and $R-j$ servers idle. If, on the contrary, j particular servers are observed, what is the probability that they will be busy, this time without specifying the state of the other $R-j$ servers?

Let us denote $H(j)$ as the probability of such an event in the following. When the above condition is imposed on the total number of busy servers, this can be written as

$$H(j) = \sum_{i \leqslant R} P_i P(j \text{ servers are busy} \mid i \text{ in total are busy})$$

There are $\binom{R}{i}$ ways of choosing the i busy servers among which $\binom{R-j}{i-j}$ realize our event. (They are listed by first placing the j servers. There then remains $i-j$ other servers to be placed in $R-j$ "slots".) Assuming that all the occupation situations are equally probable, we have

$$H(j)=\sum_{i=j}^{R} P_i \frac{\binom{R-j}{i-j}}{\binom{R}{i}} \tag{2.22}$$

Here, eqn (2.12) leads after a few manipulations to

$$H(j)=\frac{E(R, A)}{E(R-j, A)} \quad \text{(the Palm–Jacobeüs equation).} \tag{2.23}$$

The same approach can be applied to any distribution; $H(j)$ is derived from P_i. Thus, for the Bernoulli distribution with P_i given by eqn (2.21), we have

$$H(j)=\sum_{i=j}^{R} \binom{R}{i} a^i (1-a)^{R-i} \frac{\binom{R-j}{i-j}}{\binom{R}{i}}$$

i.e.

$$H(j)=a^j$$

For Engset traffic, eqn (2.22) leads to

$$H(j)=\frac{\mathrm{Eng}(N, R, \alpha)}{\mathrm{Eng}(N-j, R-j, \alpha)}, \tag{2.24}$$

$$\mathrm{Eng}(N, R, \alpha)=\frac{\binom{N-1}{R}\alpha^R}{\sum_{k=0}^{R}\binom{N-1}{k}\alpha^k}$$

Note that, in the Erlang case, eqn (2.23) for $j=1$ gives

$$H(1)=\frac{A\{1-E(R, A)\}}{R}, \tag{2.25}$$

i.e. the carried traffic per link. When j is small, $H(j)=a^j$, as for the Bernoulli traffic (a is the carried traffic per link).

Also note that the Bernoulli distribution is the limit of the Engset distribution when $R>N$. Similarly, if N tends to infinity and α to zero in eqn (2.17) in order to keep the product $N\alpha$ constant, using the Stirling formula shows that then eqn (2.17) admits eqn (2.12) as a limit with $N\alpha = A$, a result which agrees with intuition.

2.9 Waiting system

It is not necessary here to make a distinction between offered and carried traffic. Similarly, the concepts of congestion are irrelevant.

Nevertheless, waiting systems lead to important mathematical developments; we can only envisage the simplest cases.

Among these, the *M*/*M*/*R* queue is in current use. This time, the birth and death process has an infinite state space with

$$\lambda_k = \lambda \qquad k = 0, 1, 2, \ldots$$
$$\mu_k = k\mu \qquad k = 1, \ldots, R$$
$$\mu_k = R\mu \qquad k > R$$

Here it is necessary to write down a condition for the existence of steady-state conditions; condition (2.8) applied to the coefficients λ_k and μ_k is written

$$\lim\left(\frac{\lambda_k}{\mu_{k+1}}\right) = \frac{\lambda}{R\mu} < 1 \qquad (2.26)$$

By means of this assumption, the solution can be found as

$$j \leqslant R \qquad P_j = \frac{A^j}{j!} P_0$$
$$j > R \qquad P_j = \frac{A^R}{R!}\left(\frac{A}{R}\right)^{j-R} P_0 \qquad (2.27)$$

where

$$P_0 = \left\{1 + A + \frac{A^2}{2!} + \ldots + \frac{A^R}{R!}\left(1 + \frac{A}{R-A}\right)\right\}^{-1}$$

We have set $A = \lambda/\mu$; this is the offered traffic equal to the carried traffic.

The waiting probability is given by

$$P(\text{waiting}) = \sum_{j \geqslant R} P_j = \frac{R}{R-A} \frac{A^R}{R!} P_0 = E_2(R, A) \qquad (2.28)$$

which is the second Erlang formula, known as the Erlang *C* formula (see Appendix at the end of the book for the calculation method and graphs).

The use of standards for dimensioning equipment most often specifies the waiting distribution function. In fact, we attempt to meet a criterion of the type

$$P(\text{waiting time} > t_0) < p$$

For an $M/M/R$ queue, we have

$$P(\text{waiting time} \leqslant t) = 1 - E_2(R, A)\, \mathrm{e}^{-(R-A)t}$$

which solves the dimensioning problem.

Unfortunately, the majority of systems are complex, giving rise to a sophisticated mathematical treatment; in practice, the engineer will then make use of tables. This is particularly the case for the $M/D/R$ system which is so common when dimensioning the exchange is considered; the signaling transmitters–receivers etc. have an occupation time close to a constant duration (see Appendix at the end of this book).

However, we give a final formula in widespread use which provides the average of the waiting time, the Pollaczek–Kintchine formula. It relates the waiting and service times for a queue with Poisson arrivals. If we let $E(x)$ be the average service time, c be its coefficient of variation (the ratio of the standard deviation to the mean) and $\rho = \lambda E(x)$ be the "offered traffic", the waiting time (before the beginning of service) is given by

$$W = \frac{\rho E(x)}{2(1-\rho)}(1 + c^2) \tag{2.29}$$

and the sojourn time (waiting plus service) by $W + E(x)$. The variance of the waiting time involves the third moment of the service distribution:

$$E(x^3) = \int x^3\, \mathrm{d}F(x)$$

$$\operatorname{var}(W) = W^2 + \frac{\rho}{3(1-\rho)}\frac{E(x^3)}{E(x)} \tag{2.30}$$

References

This chapter should be seen as an elementary introduction to the theory and tools of "teletraffic". However, hopefully it should provide an adequate basis for reading the rest of the book, which can be seen in fact as an application (except perhaps for Chapter 9, which requires a good knowledge of queuing theory to be used fully).

We only quote textbooks at this stage; papers and articles would be too numerous and without interest for general information. References on more specific subjects can be found in the following; experimental verification of the Poisson assumption as well as service time characteristics are covered in Chapter 8.

For the reader wishing to delve more deeply into the probability aspects of the chapter (especially stochastic processes), [Feller 71] and [Cinlar 75] are especially recommended.

For the more specialized topics of this book, the reader should consult the literature on two topics: the classical literature on teletraffic such as the highly reputable *Proceedings of the International Teletraffic Congresses* and [Syski 86] (this reprint of the original 1960 edition is highly mathematical); the literature on "queuing theory" which is more recent and increasing daily) especially the classical and excellent studies [Kleinrock 75] and [Lavenberg 83].

Historical account. Teletraffic was born at the beginning of the century, with the works of the great pioneers (Erlang, Molina, Engset, Vaulot etc.). Up to the 1950s, it was the exclusive field of telephone engineers and probability theorists. The growth and development of computers and the need to evaluate their performance caused computer scientists to join in its study. We are indebted to them for the classical results on queuing networks (for the Gordon–Newell and Baskett–Chandy–Muntz–Pallacious theorems, see [Lavenberg 83]). Nowadays, teleinformatics represents the union of these fields; joint congresses and reviews are commonplace as there are common problems to be solved.

[Kleinrock 75] L. Kleinrock. *Queueing Systems*, Vols 1 and 2, John Wiley, New York, 1975 (Vol. 1), 1976 (Vol.2).

[Lavenberg 83] S. Lavenberg. *Computer Performance Evaluation Handbook*, Academic Press, New York, 1983.

[Syski 86] R. Syski. *Introduction to Congestion Theory in Telephone Systems, Studies in Telecommunications*, Vol. 4, North-Holland, Amsterdam, 1986 (reprinted from Oliver and Boyd, London, 1960).

[Feller 71] W. Feller. *An Introduction to Probability Theory and Its Applications* (2 volumes), John Wiley, New York, 1971.

[Cinlar 75] E. Cinlar. *Introduction to Stochastic Processes*, Prentice-Hall, Englewood Cliffs, New Jersey, 1975.

Chapter 3

STRUCTURE OF THE SWITCHING NETWORKS

This chapter marks the beginning of the first part of the book devoted to switching networks; it describes the components of the networks, their structure and their utilization.

Section 3.1 introduces the components; these are matrices which are assembled into a linked network. Economic reasons dictate the use of structures where it is impossible to route a call between a pair of idle inputs–outputs; this is matching loss or internal blocking, a central concept in the study of switching networks which is presented. The concept of the channel graph is also introduced; it facilitates the calculation of matching loss in the following chapter.

In Section 3.2 the use of switching networks is illustrated by describing the communication process in a subscriber exchange and the search for a route.

Finally, Section 3.3 describes the main types of architecture which are currently used.

We have placed more emphasis on what the structures have in common rather than on the technological differences which are of little importance here. However, it seemed worthwhile to devote a few lines to the technology of time-division switching in order to demonstrate its functional identity with electromechanical systems or space-division switching electronic systems.

3.1 Components

The switching network must be capable of connecting two end groups, "inputs" and "outputs". Let us consider for simplicity the case of a square network with N inputs and N outputs. As an example, N varies from 1000 to 50 000 approximately.

The ends are materialized by individual connections (analog channels) or multiplexed on a PCM support as shown in Fig. 1.2. The elementary component which connects these ends is called the switching point or contact, or crosspoint. Its physical structure can vary; it can be an electromechanical relay, electronic circuit with transistors etc. For time-division switching, it does not really exist but everything happens as if it existed in this form (see below). Sometimes, the term "relay" is used but this is not the proper terminology.

These switching points are assembled into matrices which constitute the true "atoms" of the more complex edifices actually used.

3.1.1 The basic element: the matrix

The simplest matrix is used for analog networks (the so-called space-division switching). Figure 3.1(a) represents such a system and Fig. 3.1(b) gives the customary symbolic representation.

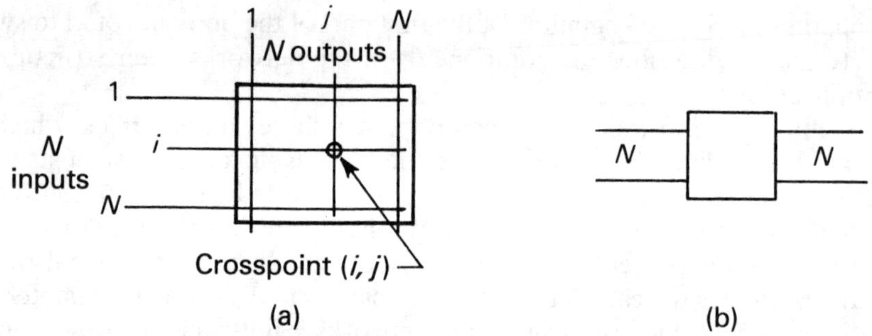

Fig. 3.1 An $N \times N$ matrix.

Each end carries two wires to be switched, and sometimes a third wire or even four wires (if the two transmission directions are separate). Then, several matrices similar to those shown in the diagram are used, and these are controlled in parallel.

The input i is connected to the output j by closing the crosspoint (or the relay (i, j)), the rule of the game consisting in only closing once for each row and column.

Such a matrix switches N inputs to N outputs, i.e. it allows each input to be connected to each output at the same time at the cost of N^2 elementary relays (the exact physical constitution of these "relays" is of little importance here). It can be understood that this solution has little practical interest in terms of the cost as soon as N becomes large; in particular, no more than N relays are used at the same time.

In fact, these matrices are used in more complex assemblies. A given technology makes it possible to realize matrices of limited size; the crossbar systems used matrices of approximately 50×50; the space-division switching electronic systems (reed relays on printed circuits or miniselectors) use matrices with $N = 4$, 8 or even 16 or 32.

Time-division switching makes it possible to achieve larger sizes and above all operates directly on the PCM multiplex (hence its interest). A distinction is made between two basic components: the time-slot interchanger (the T matrix); the time-multiplexed matrix or space division matrix S.

The time-slot interchanger (Fig. 3.2)

This element consists of n buffer memories each of which contains an entire frame (each buffer consists of 32 bytes in series or in parallel). Each multiplex (n per matrix) refreshes the contents of its buffer, each byte being thus updated every 125 μs.

The generation of the time slot of rank i for the n outgoing channels is carried out during the 3.9 μs of the slot i (in Fig. 1.2, each time interval from T_i to T_{i+1} is called a time slot.

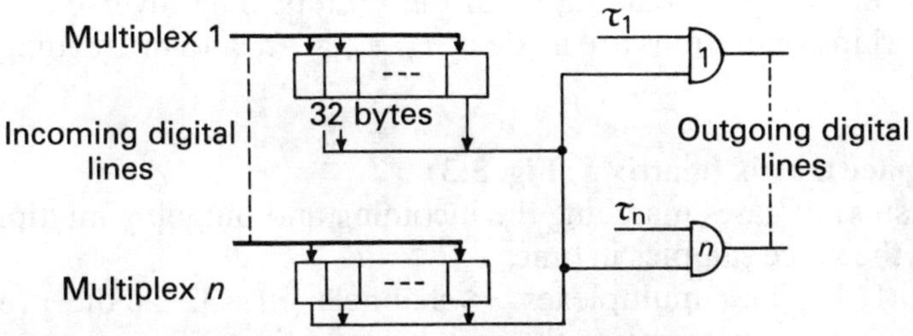

Fig. 3.2 The T switch.

- Each interval of 3.9 μs is decomposed into $n+1$ elementary times $\tau_0, \tau_1, \ldots$.
- At the time τ_0, the n writing operations are carried out simultaneously.
- At the time τ_1 of the interval i, a byte is read (among the $32n$ possible bytes) and directed to output 1. It constitutes the contents of the channel i of the outgoing multiplex 1.
- At time τ_k ($k \leqslant n$) of the same interval, a byte is read and directed to the outgoing multiplex k.

Thus, during each interval of 3.9 μs, a channel will have been read on each incoming multiplex constituting the interval i of the outgoing multiplex.

In addition, the switch consists of buffer memories belonging to a control memory containing the address of the multiplex to be read during the time τ_k of the interval i and an output gate reconstituting the PCM multiplex.

Such a device makes it possible to relate each channel of each incoming multiplex to each channel of each outgoing multiplex; the channel j of the incoming PCM p is put into contact with channel k of the PCM q by writing the address (j, p) at time τ_q of the interval k. The switch T described here is therefore the functional equivalent of a matrix with $32n$ inputs and $32n$ outputs (the matrix of Fig. 3.1 with $N=32n$).*

The limitations to the size of the matrix T derive naturally from the speed of the reading operations ($n+1$ in 3.9 μs). At present, matrices are produced accepting 32 incoming PCMs; each interval τ_i lasts 118 ns. Obviously, each switch is preceded by a serial–parallel transformation to reduce the necessary speed.

The number of incoming and outgoing channels may differ: in particular, matrices are produced with 32 rows of 32 channels (1024 incoming circuits) and with 16 rows of 64 outgoing channels. These lines operate at a rate which is twice as fast; all operations take place as if lines 0 and 1, 2 and 3 etc. in Fig. 3.2 were being multiplexed. This does not change the equivalent diagram but divides the size of the associated matrix S by 2 (see below).

It is said that this structure is under "output control"; the switching is carried out after writing into the input buffer and the entire outgoing time slot i is produced at time i by reading in all the buffers. The inverse structure also exists, the writing operations are made at $\tau_0, \tau_1, \ldots, \tau_n$ and the reading operation at τ_{n+1}.

The space matrix (matrix *S*, Fig. 3.3)

This is a system of gates matching the incoming and outgoing multiplexes without shifting the voice samples in time.

Each of the lines multiplexes m channels ($m=32$ at the present time; sometimes $m=64$ is used inside the switches). At times T_k the control memory of each outgoing link opens one of the gates transmitting the sample of rank k from an incoming link to the outgoing link. At each successive time T_k, a differ-

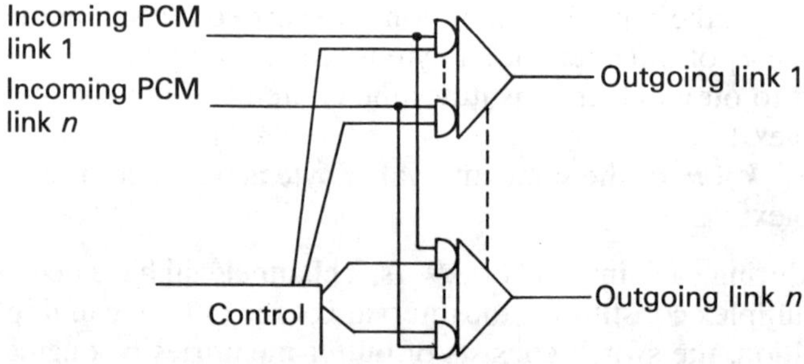

Fig. 3.3 The space switch.

*At least for the basic Conference of European Postal and Telecommunications (CEPT) PCM system (32 channels).

ent gate will open providing a different configuration, i.e. an interconnection between the incoming and outgoing links; the equivalent circuit diagram of the space switch consists of m matrices of size $n \times n$.

Both the advantages of the structure and its technological limitation can be seen here; a single circuit is equivalent to m matrices, the disadvantages being the switching speed and the capacity of the control memory.

As distinct from the time-slot interchanger, the space matrix processes each byte serially.

The time-division technique makes it possible to reach up to 1024 inputs.

$N \times M$ matrices ($M < N$) are used to concentrate the traffic. The representation in Fig. 3.1(b) will be used irrespective of which technology (space division or time division) is used.

Note that, instead of T and S matrices, they are sometimes called M (memory) and G (gate) matrices respectively.

3.1.2 Assembly into complex networks

When the range of selection increases, it is obviously not possible to let the size of the matrix increase indefinitely. The elementary matrices are then assembled into networks; we shall give a summary of the structures which are in current use today.

The first scheme consists of using two "stages" of matrices according to Fig. 3.4. The link system has been invented; each square represents a matrix

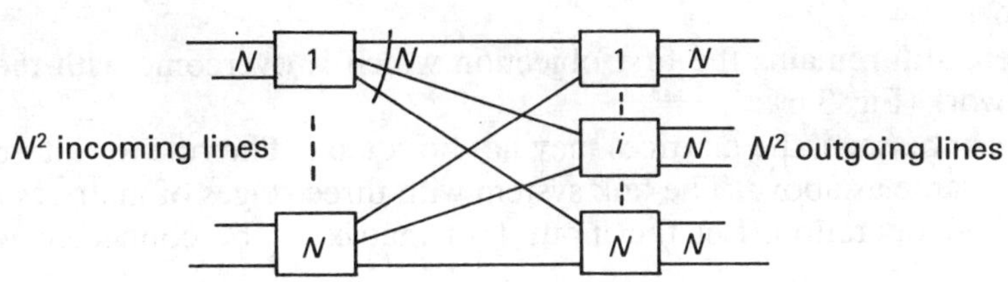

Fig. 3.4 Link system with two stages of matrices.

similar to that of Fig. 3.1; the ith matrix of the first stage has its outputs connected to the matrices of the second stage according to the following rule.

Output j of the matrix i is connected to the input i of the matrix j.

This makes it possible to connect N^2 inputs to N^2 outputs by using $2N$ matrices of size $N \times N$, each with N^2 contact points, i.e. $2N^3$ contacts instead of N^4.

The network constructed in this way cannot be used in this state; in particular, the following should be noted

- The absence of mixing ("distribution") between the traffic flows, i.e. the calls going from i to j take the link ij and do not mix with the other flows. The importance of distribution will be seen later (see Chapters 4 and 5).
- Limited access, i.e. a single link (a single channel) and therefore a single route between our matrices i and j. It is therefore only possible to carry a single communication at the same time between the incoming block i and the outgoing block j. In fact, with time-division techniques, a derived structure is used where each link becomes a multiplex, thus giving f routes between each pair of matrices (Fig. 3.5).

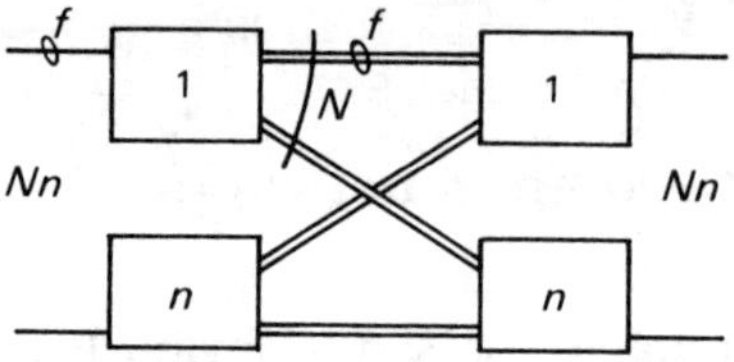

Fig. 3.5 A two-stage network using n $N \times N$ matrices for each stage. $N = nf$.

Some time-division networks (such as the UXD system of British Telecom, and System X with a low capacity or the DMS system of Northern Telecom) use the structure in Fig. 3.5. The set of links between each pair is a PCM multiplex and $f = 32$ or 64.

There still remains the first objection which is overcome with the three-stage network (Fig. 3.6).

As can be seen, the matrices may not be square. The rule for interconnection is the same as above. The link system with three stages of matrices ensures the required operation. Let the input 1 of matrix A_1 be connected with the

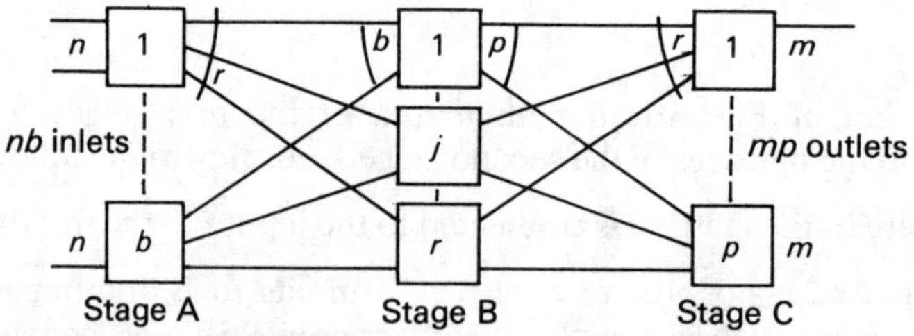

Fig. 3.6 A three-stage network.

output k of matrix C_p of the third stage. Choosing $j(1 \leqslant j \leqslant r)$, the contact $(1, j)$ of the first matrix, the contact $(1, p)$ of the matrix B_j of the middle stage and the contact (j, k) of the matrix C_p on the third stage are switched on. In the above example, each input–output pair may be connected by r different routes (r possible choices for the central matrix). The traffic flows from the matrix A_1 to the matrix C_p are mixed with the flows from A_1 to the other matrices at the first stage of the links. Similarly, all the flows to C_p are mixed at the second stage; distribution actually takes place.

The number of necessary "relays" will depend on the size of the matrices; *nb* inlets are connected to *mp* outlets by the following number of connection points:

$$nrb + brp + mrp$$

For a symmetric network, $n = m$ and $b = p$; this network provides the functions of the square matrix as soon as $r > 2n - 1$ (see the next section); the structure defined in this way is called the Clos network. In this case the number of necessary relays is equal to $b(2n-1)(2n+b)$ for *nb* inputs. The optimum growth then leads to a variation in $N^{3/2}$ ($N = nb$); the Clos network is said to increase as $O(N^{3/2})$ and the matrix as $O(N^2)$.

As for the two-stage networks, the structure is sometimes used with the internal links replaced by multiplexes. This is so for the switching network of a large number of systems, e.g. UT 100/60 (System Proteo 2 of Italtel); this is a network of the so-called *TTT* structure (Fig. 3.7).

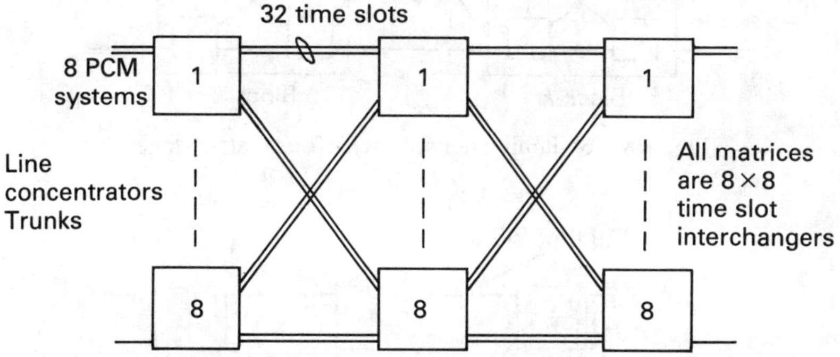

Fig. 3.7 The three-stage switching network UT 100/60 (a "switch section").

The figure shows the basic "switch section". The actual network has up to 16 "planes" each made up of two sections. The interconnection system between concentrators or trunks and the network is of prime importance for such networks where no mixing exists between "sections".

Using the same model, it is possible to create networks with four or more matrix stages which of course increases the distribution and the number of different routes between the ends.

It is also possible to create structures using another linking scheme. The performances of a network will vary according to its structure (see Section 3.4).

Figure 3.8 gives a rather common example of a structure with four stages.

For time-division networks, of the functional structures shown in Figs 3.6, 3.7 or 3.8 is applicable is not always immediately apparent.

For example, consider the "medium-capacity" version of the structure of the EWSD switching network. It is a three-stage network, of the type in Fig. 3.8, the central matrix being a space matrix. The network is called a "*TST*" type (Fig. 3.9).

The basic time-slot interchanger matrix switches four multiplexes of 128 time slots each (the rate of the internal PCM multiplex is thus 8 Mbits/s). The space switch performs 16 × 16 interconnections. The networks therefore switch

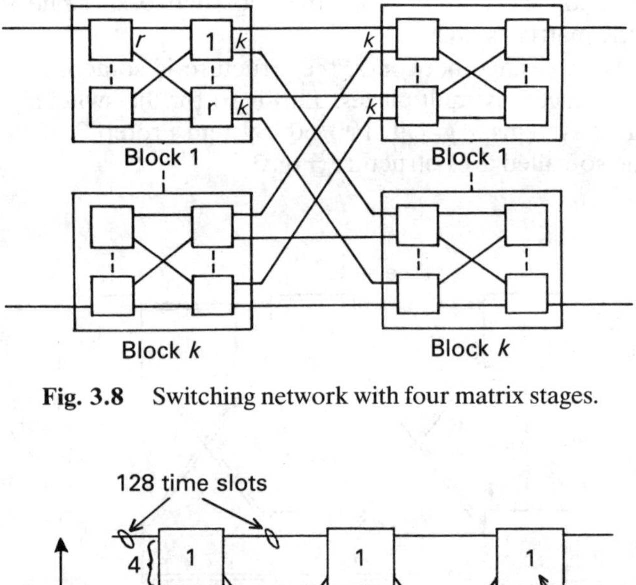

Fig. 3.8 Switching network with four matrix stages.

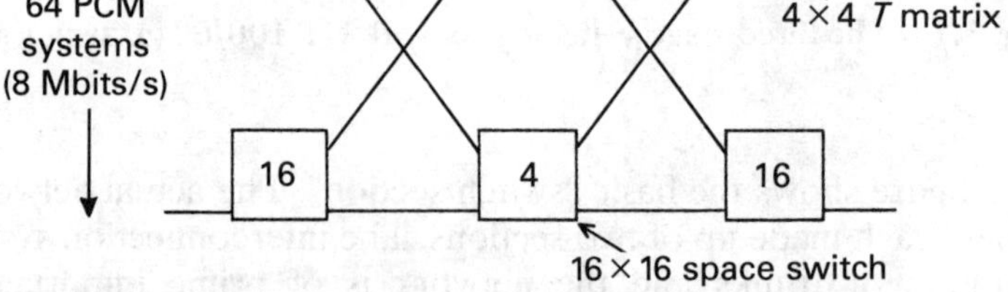

Fig. 3.9 The *TST* EWSD switching network (medium capacity).

64 × 8 Mbits/s systems, i.e. 16 000 channels. A higher capacity is obtained by expanding each space switch into a three-stage subnetwork, leading to 128 internal multiplexes.

For reliability reasons, the network is duplicated; two identical switching systems work in microsynchronism (see Chapter 7). A single network is able to process the whole traffic. Other systems (such as ESS4 or E10) use a load-sharing mechanism, a solution which presents the advantage of doubling the available capacity during the no-failure periods.

Figure 3.9 shows the actual (hardware) wiring. Functionally, however, each space switch is able to produce a distinct interconnection pattern at each successive time slot. This means that, from the viewpoint of paths of 64 kbits/s, a single space switch is equivalent to 128 replications of a single 16 × 16 switch. In this equivalence, the 128 time-slot internal multiplex turns into 128 distinct links, each of which is dedicated to the corresponding space switch replica.

Finally, an equivalent "space" representation is provided by Fig. 3.10. The value of this equivalent representation is to shed light on the functional structure and to demonstrate the distributions which actually exist and the set of routes. The concept of the channel graph (see Section 3.1.5) plays an identical role.

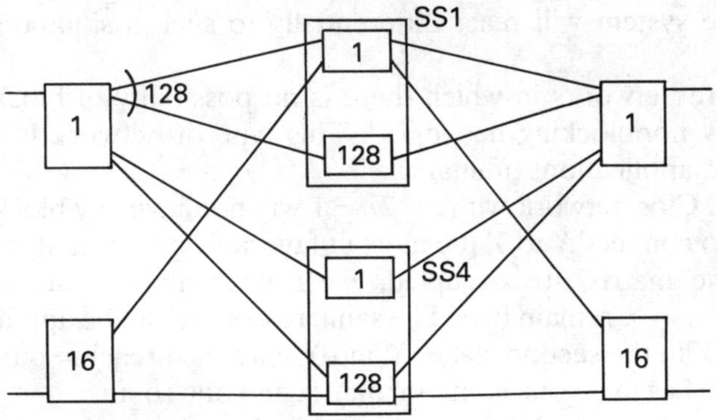

Fig. 3.10 The functional structure of the EWSD network in Fig. 3.9.

3.1.3 Matching loss

The elementary matrix in Fig. 3.1 is simple to handle. In contrast, the reduction in the number of crosspoints of the link network makes it necessary to complicate the control function. In fact, it is necessary to carry out a path search (finding a matrix j in Fig. 3.6 where the input links $(1, j)$ and output links (j, p) are free at the same time). This search may end in failure. This is called matching loss or internal blocking.

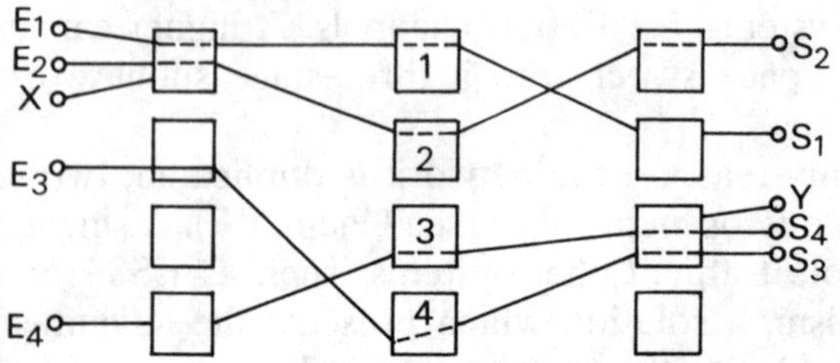

Fig. 3.11 Matching loss (only busy links are represented).

This phenomenon lies at the heart of traffic problems in switching networks and understanding this basic concept is of prime importance. Let us therefore consider a three-stage network similar to that in Fig. 3.6 with $r=n=4$ (the number of stages and sizes of the matrices can be varied without changing the phenomenon (Fig. 3.11).

Let us attempt to connect the X and Y outlets. At the output of the X matrix, the connections to matrices 3 and 4 of the central stage are idle but those of these matrices to the output are occupied; the two idle ends cannot be connected. This is the situation of matching loss. According to the implemented operation, the system will react differentially to such a situation (see Section 3.2.2).

There are networks in which there is no possibility of blocking; they are called "strictly nonblocking networks". This type of network is only used for very particular applications (military).

Thus the Clos network with $r=2n-1$ will not have any blocking as can be easily seen. To connect X to Y, the most unfavorable case is that, when the $n-1$ sources of the matrix are occupied, $n-1$ internal links are occupied and $(2n-1)-(n-1)=n$ remain free. The same reasoning can be applied to the connection from Y to the second stage. X and Y may each reach n out of the $2n-1$ matrices; there has to be one common matrix and one route.

The phrase "nonblocking networks in the broad sense" is used if particular routing rules make it possible to avoid the blocking, e.g. rearrangements (see Section 3.2.2).

However, most networks have blocking; for these, it is necessary to know how to evaluate the blocking probability (or congestion). The question of why blocking networks which pose such problems are used can be asked. The reason is economic; a telephone network cannot avoid the phenomenon of rejection of communications for obvious reasons linked to the random nature of the demand. Consequently, it is important to design homogeneous networks in which the probabilities of rejection are distributed. One of the sources of this rejection lies in the intercenter trunk groups. There the loss is limited to values close to 1% in general. Congestion in the network is therefore limited to compatible values (10^{-3}–10^{-2} in most standards).

3.1.4 Composition of the networks

Current technological constraints dictate a maximum value and for economic reasons even a recommended value for the parameter n (the size of the elementary matrix) which governs the selection range of n^2 inputs and outputs for a three-stage network using all the matrices of size n. In order to achieve larger network sizes, imperfect linking is used obtained in practice by assembling the basic building blocks made from networks with two, three or four stages.

Furthermore, a network will have to develop by extensions spread over time. The final capacity will not be installed directly; hence the interest in modular structures with the potential for simple and economic growth.

For such structures the crucial point is the choice of interconnection rules leading to regular architectures and to feasible extensions. The "main distribution frame" is responsible for these interconnections. The references should be consulted for extra information on the methods used. There are also some references describing effective implementations which illustrate these rules. The reader is advised to refer to these descriptions which complete the information in the preceding section.

Note that the availability constraints often make it necessary to install two networks, one being capable of handling the entire traffic by itself. This redundancy can be exploited to facilitate the extension operation, by disabling the half to be extended.

3.1.5 The channel graph

It is worth visualizing the total number of routes possible between one input and one output in a switching network; the complexity of the graph obtained in this way (called the channel graph) gives an idea of the distribution and the quality of the network. In addition, blocking calculations are based on this graph.

For example, the three-stage network in Fig. 3.6 has the channel graph shown in Fig. 3.12(a). The channel graph of the network in Fig. 3.7 is shown in Fig. 3.12(b) and that in Fig. 3.8 is shown in Fig. 3.12(c).

The construction of the graph in Fig. 3.12(a) is obvious. Let us choose an input line and an output line; each route between these ends consists of two links in cascade which makes use of one of the r central matrices. In the case in Fig. 3.12(b) the connection between the matrices provides f equivalent channels with f branches in parallel.

Similarly, in the network in Fig. 3.8, a route is determined by a matrix of the second stage (r routes in total) and crosses the two central stages.

The symmetry of the structures makes it unnecessary in general to consider as many graphs as input–output pairs.

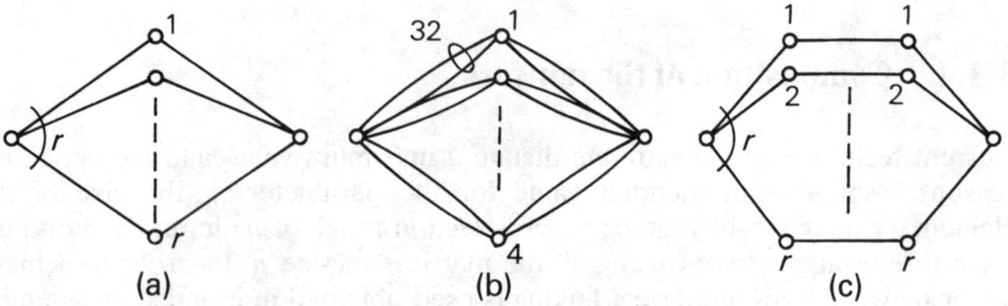

Fig. 3.12 Channel graphs (a) for Fig. 3.6, (b) for Fig. 3.7 and (c) for Fig. 3.8.

The routes of the networks in Figs 3.6–3.8 are all separate; this is referred to as series–parallel networks and graphs.

For nonseries–parallel networks (Fig. 3.13), the channel graph (Fig. 3.14(a)) takes a different form, certain links being common to several routes. Sometimes, a canonical graph (Fig. 3.14(b)) is then defined where each route has its own distinct series branch; it is then necessary to reproduce certain links several times.

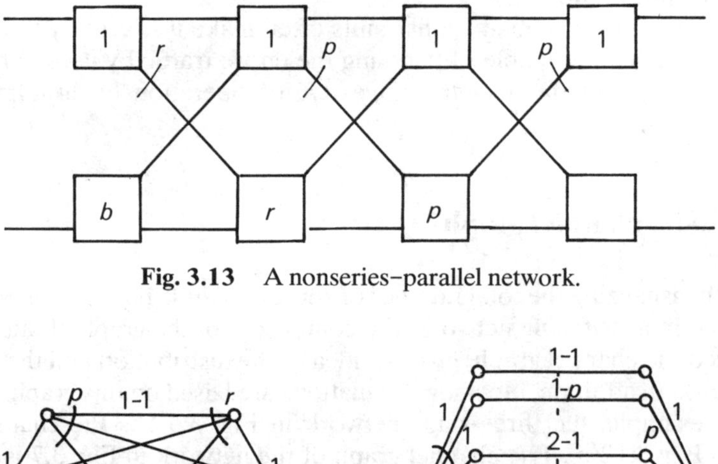

Fig. 3.13 A nonseries–parallel network.

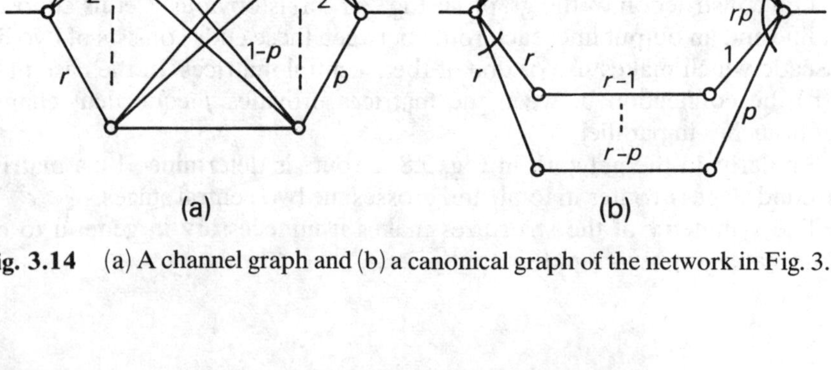

Fig. 3.14 (a) A channel graph and (b) a canonical graph of the network in Fig. 3.13.

In the general case the canonical graph consists of M branches composed of s arcs in series (where s is the number of stages of the network) and which represent all the possible routes differing by at least one link. The use of this graph will be seen in the next chapter.

A channel graph can have a fairly involved structure. As an example, let us construct the channel graph for the ESS4 time-division system; ESS4 is the high-capacity toll exchange developed by the Bell System.

The external stages of the network are 128 incoming and 128 outgoing "switching and permuting circuits", each of which is constructed from eight time-slot interchanges and eight S matrices, of size 8×8. The interconnection makes use of 128 time-slot multiplexers.

The central stage is made of four "time multiplex switches". They are two-stage switching arrays, composed of identical 16×16 S matrices. Each S matrix of the switching and permuting circuit has two links (two multiplexes) to each time multiplex switch. Figure 3.15 shows the ESS4 network.

The network is said to be of the type TS^4T. Bearing in mind that the internal multiplexes have 128 time slots, we expand the S stages as in Fig. 3.10 in order to provide the channel graph (Fig. 3.16).

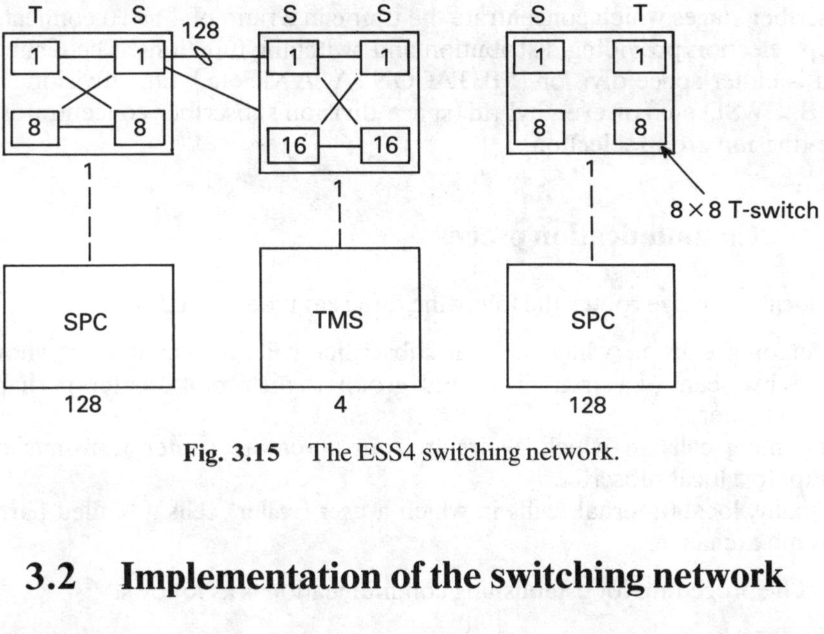

Fig. 3.15 The ESS4 switching network.

3.2 Implementation of the switching network

A complete description of the functioning of a network depends on the structure and the type of center considered (see for example the descriptions given in the references). The most interesting points for the flow of traffic are the call flow diagram (chaining of the phases) and the path-searching algorithms.

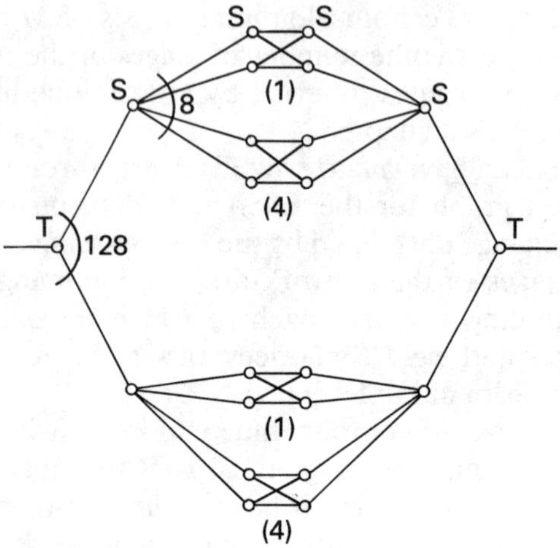

Fig. 3.16 The channel graph of the ESS4 network.

As an example, we shall consider a local exchange (an exchange whose size varies between 15 000 and 60 000 lines). Such an exchange usually consists of subscriber stages which concentrate the traffic in a ratio of 4 to 10 connected by group selection providing distribution and switching functions. The technology used is either space division (METACONTA, AXE etc.), time division (ESS4, E10B, EWSD etc.) or even hybrid (space-division subscriber concentrators and time-division group selection).

3.2.1 Communication process

Our local exchange routes the following different types of call.

- Outgoing calls in which a local subscriber calls a distant user whom he reaches by means of a circuit (in a trunk group) to the remote center (perhaps via a transit center).
- Incoming calls in which a circuit from a remote center transmits a call attempt to a local subscriber.
- Finally, local (internal) calls in which a user (caller) calls a "called party" of the same exchange.

The procedure for establishing communication is as follows.

Outgoing call

(1) Preselection. When the fact that the caller has unhooked is detected, the control system looks for specialized equipment (dialing receiver) which is responsible for supplying the subscriber (space systems), sending the tone and

often receiving the digits. It is necessary to connect this equipment to the subscriber across the network or a part of the network. The control system must carry out a path search and send a connection order to the marker. Marker is the term given to the peripheral responsible for controlling the network; its implementation varies with the technology (space division or time division) (see [Grinsec 83]).

(2) Dialing. The dialing receiver collects and transmits successive digits to the control computer. Once a sufficient number of digits has been received (two to four), the next stage can commence.

(3) Selection. As soon as it is possible to determine the direction of the call, it becomes necessary to find an idle circuit in this direction and a path between the caller and this circuit (the term junction is used in analog technology). It is also necessary to transmit signaling to the destination exchange, i.e. all the service information concerning the call (the type of call, number requested, billing information etc.). A signaling sender is usually responsible for this task. It is therefore necessary to find such equipment. Signaling information may use either the same path as the speech signal ("channel-associated signaling") or a separate path ("common-channel signaling"). In fact, channel-associated signaling actually uses the same physical path only for analog transmission ("in-band" signaling). For digital transmission the frame carries the signaling of its channels; for instance, in the CEPT PCM, time slot 16 is dedicated to the signaling of the 30 channels. In such cases a permanent connection is provided between it and the signaling device or the common control which can take its place. The final case of common-channel signaling consists of a data transmission network which is separate from the telephone network. The senders, are then organized in a separate processing unit. There is obviously then no path search. The increase in transfer speed and its improved signaling capabilities explain the interest in this solution (see also Chapter 11).

(4) Conversation. At the end of transmission and the acknowledgment of positive reception, the equipment concerned and the paths are cleared and the call passes to ringing and then to conversation.

(5) Clearing. When one of the two communicators hangs up, the equipment and path are cleared in memory (updating of the tables) and in the network (sending an order to the markers for disconnection) and then billing.

The sequence of operations is summarized in Fig. 3.17.

It should be noted that in this diagram there is overlapping of the phases which is translated by an overlapping of the occupation of the paths and the equipment involved.

Incoming calls

(1) Preselection. A call coming from the outside is detected by the center (detection of a suitable code on the common channel or of a signal on the incoming junction; low-frequency analog circuit termination equipment). For

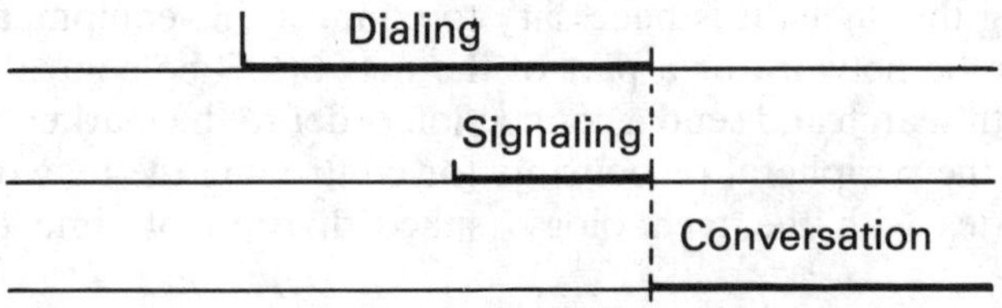

Fig. 3.17 Establishment of an outgoing call.

transmission within the band, the incoming junction is connected to a signaling receiver which decodes the called number.

(2) Selection. The control system tests the state of the called subscriber and then possibly looks for a path connecting it to the incoming circuit.

(3) Conversation. As previously.

(4) Clearing. As previously.

The diagram for the sequence of operations is shown in Fig. 3.18.

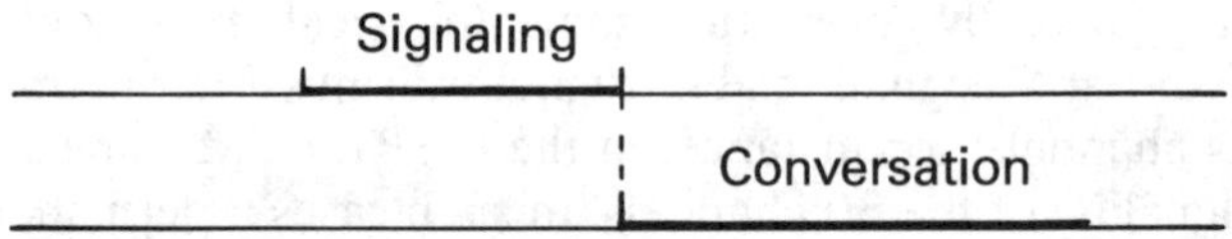

Fig. 3.18 Establishment of an incoming call.

Local calls

It is not difficult for the reader to imagine the sequence of operations by combining the incoming and outgoing sequences.

Note that a large number of variations are possible on the above theme. Thus, the connection with the dialing receiver equipment may be made only via the subscriber stages (e.g. the subscriber line switching units in the AXE system). The reception of dialing may even be carried out directly on the subscriber's equipment and therefore without a path search etc.

3.2.2 Path searching

In general, the establishment of a communication will therefore require several paths to be made. From the viewpoint of a traffic flow phenomena, in particular in terms of matching loss, it is necessary to distinguish between two types of search.

- Search of a path between two specified ends (e.g. selection of an incoming call), which is called "point-to-point path searching" (or selection).

- Search of a path between one end and a set of outputs (e.g. outgoing selection; the subscriber must be connected to any one of the circuits in the required direction), which is called "point-to-group path searching".

In practice, the control system carries out the path searches in memory. The central memory contains a mapping of the state of the network and the search is performed on this mapping. A point-to-group search most often proceeds as a succession of point-to-point searches. In fact, it is always possible, if required, to make the two types of search homogeneous; by adding a virtual stage the point-to-group search is converted to a point-to-point search. The additional virtual stage receives the "group" considered at the input and a single virtual point at the output.

The channel graph in Section 3.1.4. defines all the paths covered by the point-to-point selection.

In terms of the traffic performance which concerns us here, the point-to-point search is the simplest elementary operation. It will be more difficult to evaluate the performances of a point-to-group search which depend on the number of components of the group. In practice, the probability of failure will be lower than that of the preceding case; this will therefore limit the traffic which can be admitted.

The memory preserves the state of the switching network, each link being represented there by a binary element. The method of storing the state as well as the searching algorithm which is intimately connected to it are important factors governing the performances of the control system (the time taken for the search and the management of the memory). The principle will be presented using the example of the three-stage symmetric network (Fig. 3.19).

A "word" of convenient size (byte, half-word etc.) contains the state of the output links of each matrix of the first stage (a total of b words).

The binary element of rank j $(0 \leqslant j \leqslant r-1)$ of the word n_1 gives the state of the link AB connecting the matrix n_1 to the matrix j of the central stage (0 if the link is busy and 1 if it is idle).

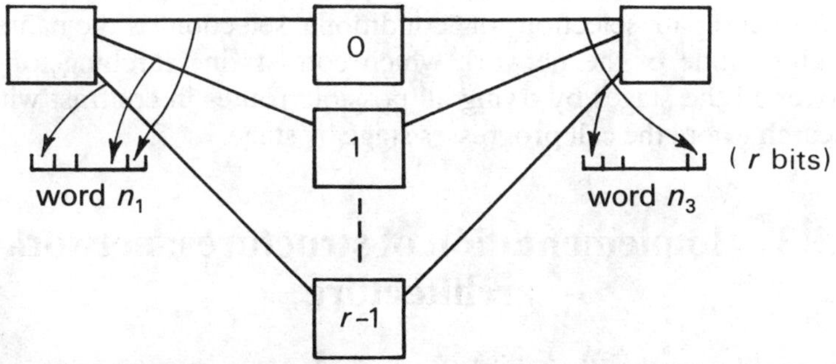

Fig. 3.19 Path search in the three-stage network.

Similarly, b "words" will contain the state of the links BC of each matrix C, Let us look for a path between the marked ends. The logical "and" of these two words will give the set of possible paths, i.e. the pair of links AB–BC terminating at the same matrix B:

$$\begin{array}{ll} \text{word } n_1 & 00101101 \\ \text{word } n_2 & 01110101 \\ n_1 \wedge n_2 & 00100101 \end{array}$$

In this example the central matrices 0, 2 and 5 can be used.

Larger networks require more complex algorithms (the complexity grows with the number of stages).

The use of computers means that a preferential choice of binary structures has to be made, i.e. matrices of size 2^n. A binary notation facilitates the representation and specification of these algorithms.

The choice of a path among those available is a characteristic of the algorithm; most current systems (if not all) use a random choice which has the advantage of distributing the wear of the equipment (a valid argument in space-division technology) and guarding against the effect of undetected breakdowns. The random choice is made by a set of cyclic pointers. Other types of choice exist and these will be described in Chapter 6.

As we have already pointed out, it may be impossible to connect two free ends (matching loss); the probability of such an event constitutes the main performance of the network and the associated algorithm (indeed, this performance depends on the implemented strategy).

When the path-searching algorithm encounters a situation of matching loss, several strategies are available to it. It can, for example, simply reject the call (the so-called loss system, which is the most common). A waiting system can equally be envisaged where the call is placed in reserve for a subsequent attempt. It is also possible to displace certain communications in progress in order to insert the new one. This is rearrangement, which produces no blocking in our network when $m = n$. If we return to the example in Section 3.1 (Fig. 3.12), the central matrix 3 would be made to take the connection $E1$–$S1$, releasing matrix 1 for the connection X–Y.

"End-to-end path selection" or "conditional selection" is the name given to the operating mode of the network which consists in searching for the path globally over all the stages by trying all possible routes in contrast with a step-by-step search where the call progresses stage by stage.

3.3 Implementation of structures; network architecture

Existing networks can be distinguished on the basis of several criteria (this is a non-exhaustive list).

- According to the technology used (space division or time division); even if the technologies produce functionally equivalent structures, the differences in the sizes of the matrices produce different behaviors in the corresponding networks.
- According to the function of the switch (subscriber concentrator, local center or transit center).
- Depending on the type of transmission (two wires or four wires).
- According to the method of connecting the ends (one-way or two-way networks).

The last two criteria require some explanation.

Type of switching

Most local subscriber centers switch two wires; the two transmission directions of the electric signal (caller–called party) are carried by the same pair.

When it is desired to amplify the signal (long-distance lines), it is necessary to separate the two transmission directions. A pair of wires is then used per direction; this is the four-wire technique. Similarly a digital (PCM) line consists of two pairs each corresponding to one transmission direction. The switch must then handle the four wires of the line simultaneously which requires four contacts per connection point. This amounts to installing two identical networks, wired and controlled in parallel. Another solution involves the use of folded networks.

One- or two-sided networks

In a two-sided network the sources of calls to be linked are partitioned into two sets, each of which is assigned a dedicated side of the switching network. The establishment of local connections requires "local junctions" which present two inputs to the network.

Figure 3.20 represents a bidirectional two-sided network (this is in fact the METACONTA 10C network). The subscribers occupy the left-hand side and the incoming or outgoing circuits the right-hand side. The dialing receivers and the signaling transmitters–receivers which were dealt with in Section 3.2 can also be seen.

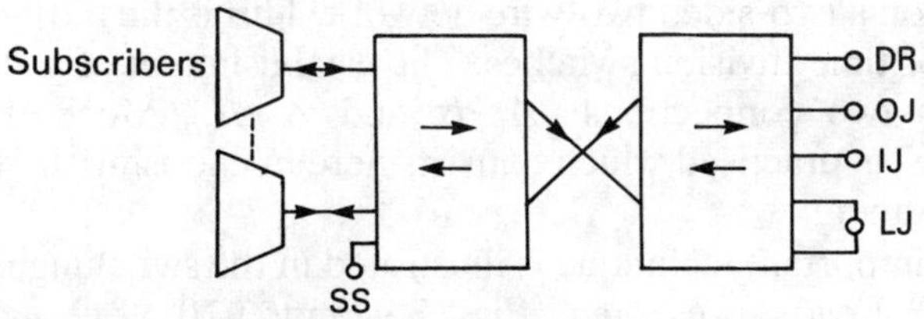

Fig. 3.20 The METACONTA, a two-sided bidirectional network.

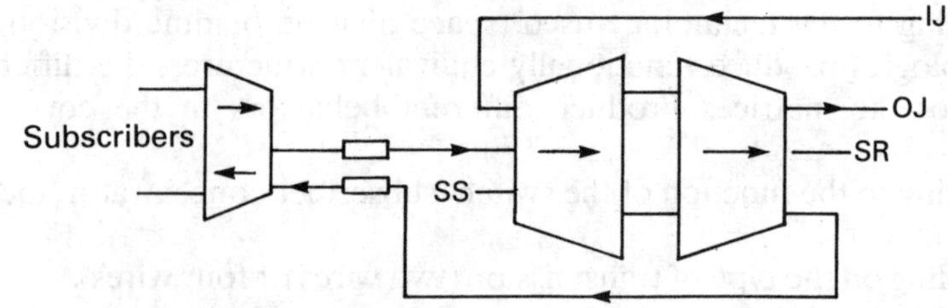

Fig. 3.21 The AXE system, a unidirectional two-sided network.

Figure 3.21 provides another example of a two-sided network, the so-called unidirectional network; the traffic flows all run from left to right. This is the space-division version of the AXE system.

These networks may exist in two- or four-wire technology (time division or space division).

In one-sided networks, the sources (circuits and subscribers) all occupy the same side. Figure 3.22 provides an example that is used in space-division technology (ESS3, EWS etc.). It may have two or four wires and is called a "reversed network".

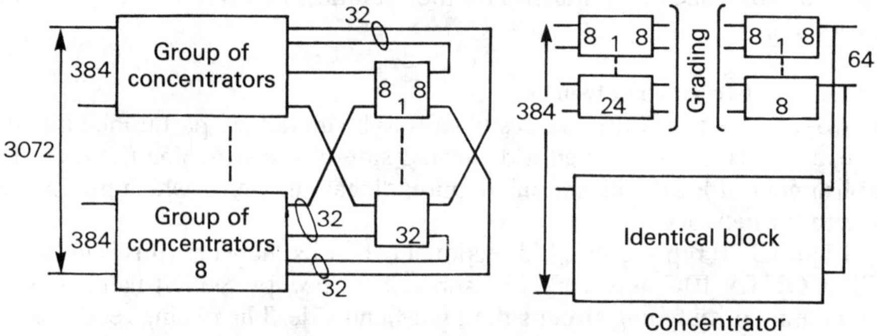

Fig. 3.22 A reversed network (ESS3 system).

Figure 3.23 illustrates the example of the four-wire one-sided "folded" network based on a two-sided two-wire network. This is the most common solution adopted for time-division switches. The establishment of a A–B communication requires two connections (A_e–B_r and A_r–B_e). Most often, so-called "control folding" is practised which consists here in choosing the two paths in a symmetrical manner.

This very important technique is illustrated in the switching network of the EWSD system already presented (Figs 3.9 and 3.10). Let us for example establish a path between a caller i and a called party j (and the return channel

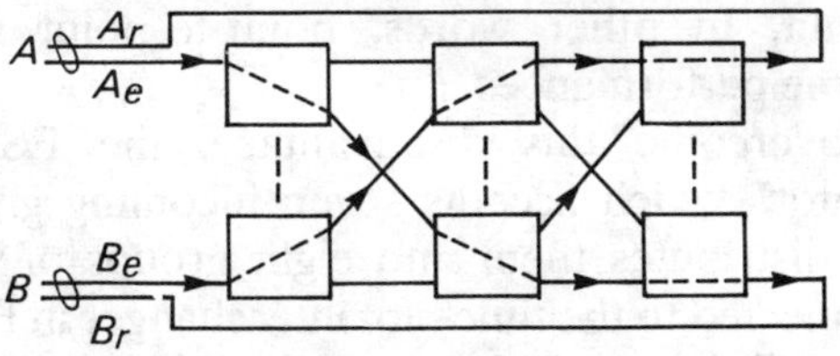

Fig. 3.23 Folded network (of the EWSD, ESS4 etc.).

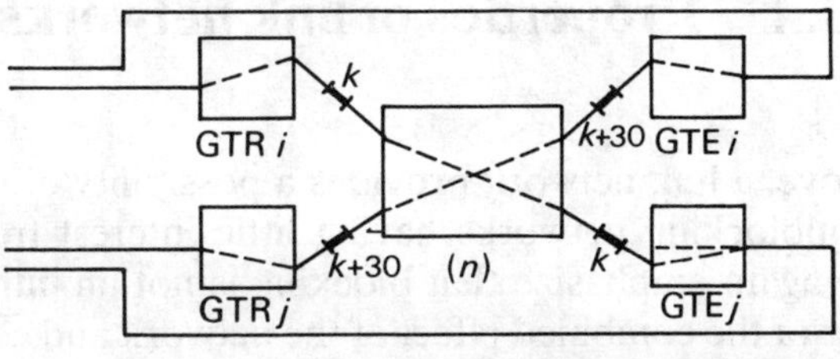

Fig. 3.24 Choice of a path in a folded network.

from j to i); if the path $i \to j$ uses the internal time-slot number k, the path $j \to i$ will use the same time slot (Fig. 3.24).

Since the space-division matrix does not displace the channels in time, the time slots have the same numbers on the right-hand side (S to T matrices).

It is easy to see that the systematic use of this rule implies that the path $j \to i$ is free as soon as the path $i \to j$ is free.

Instead of using the same time slot, any rigid assignment can be used. For instance, the rule $k \to k+64 \bmod 128$ is often imposed in order to be able to establish a connection from a time-slot interchanger to itself since then the two connections time-slot interchanger$-S$ matrix use the same multiplex.

Connection of trunk groups

In space-division technology with analog circuits, the operator will have no constraint in "splitting" a trunk group over the maximum number of matrices which will have two beneficial effects.

- It will increase the field of selection of the point-to-group search.
- It will decorrelate the different sources supplying a matrix, thereby authorizing the use of the Bernoulli distribution instead of the Erlang distribution (cf. Chapter 4).

In the PCM technique a trunk group is in the form of a pair of wires which have to be connected as a block (e.g. to a T matrix). A poor utilization of the trunk group and of the network results; for matching loss, there is no point in trying another circuit. In other words, point-to-point and point-to-group searches have the same performances.

In order to overcome this disadvantage, the ESS4 system uses a "deloader–decorrelator" which accepts seven incoming groups of 120 multiplexed circuits and distributes them into eight groups of 105 circuits. These groups are those connected to the time-slot interchanger in Fig. 3.15. Thus, each original trunk group will be connected to eight different matrices. The exceptional performances of the network owe much to this equipment [Cieslak 79].

3.4 Properties of link networks

Matching loss

As mentioned above, a link network provides a possibility of matching loss (cf. Section 3.1.3), nonblocking networks having little interest from an economic point of view. We again emphasize that blocking is not an intrinsic fault of the network but rather of the combined effect of the network and control algorithm.

A good switching network will have a low matching loss, i.e. less than the specifications in force (see Chapter 1); these usually specify blocking probabilities of between 10^{-2} and 10^{-3}.

Efficient distribution

The calculations of blocking are usually made on the basis of assumptions of symmetry and homogeneity of the traffic flows. These are favorable assumptions for traffic flow. The network must create or maintain this homogeneity by mixing distinct traffic streams. A good network must have performances (blocking probability) independent of the source-symmetry assumptions (see Chapter 5).

Reliability; resistance to breakdown

By the same token, a switching network is required to cope with one or even several breakdowns. In a link network it is necessary to distinguish between two types of breakdown: "relay breakdown", i.e. breakdown of the basic element; "block breakdown", i.e. breakdown of the multistage matrix or module. The breakdown of a crosspoint will rarely be critical; it removes one path out of a large number. The breakdown of an entire module may assume considerable significance in a time-division system where the matrices can reach large sizes (more than 1000 inputs). For this reason, most time-division networks are duplicated.

Modularity

The network consists of modules such as those described in Section 3.1.2 which are assembled by means of distributing frames. These modules come in the form of boards and frames constructed and assembled in the factory; the interconnection is made on site. It is important that most of the work is carried out in the factory (less cost and better reliability).

It is also important that extensions are as easy as possible. Highly linked structures with low blocking do not meet this criterion very well; structures with a large number of stages generally facilitate extensions.

Cost of the network

The cost of the network varies in direct proportion to the number of crosspoints for obvious reasons (the cost of the crosspoints, the supplies, the frames etc.) and in proportion to the complexity of the software; a large number of stages increases this complexity. The control system is involved in this cost through the path searching and corresponding storage in memory.

The value of a network minimizing the number of crosspoints can be understood.

Simplicity of the structure

A computer has to control the network. It will improve its performance by choosing the simplest possible structure. From this viewpoint, irregular wiring and gradings should be avoided.

The computer functions in binary mode, the use of powers of 2 optimizes the programs and memory and simplifies the handling of tables and algorithms.

As is to be expected, these qualities are more or less present in implementations. Thus, good modularity will facilitate good reliability and will lead to simple structures and easy extensions. However, the distribution will suffer and with it the blocking probability.

It is difficult to specify the best structure; this depends on the range of planned traffic, and the state of the technology influencing the relative costs. Some points of comparison will be found in [Rothmaier 81] and [Lotze 79].

References

Numerous topics have been omitted. For the very important technological aspects, which often explain the limits of a given structure, consult [Grinsec 83] or [Inose 81] (especially for digital systems). Readers interested in electromechanical or step-by-step systems, as well as gradings, can consult [Bear 76] or [Syski 86].

Many connecting networks have been described in detail; no better advice can be given than to complete this chapter by reading some of the following: [Irland 74], for ESS2 and ESS3, [Cieslak 79] for ESS4, [ITT1240 81, 85] for ITT1240 of the International Telephone and Telegraph Corporation (ITT), [Grinsec 83] for AXE, METACONTA, MT20 and E10 etc.

The specialized congresses always contain much information on systems under development (see in particular the *Proceedings of the International Switching Symposium* held in Florence in May 1984, for EWSD, D70 etc.).

The history of digital switching is described in [Joel 79].

[Bear 76] D. BEAR. *Principles of Telecommunication Traffic Engineering,* Peter Peregrinus, London, 1976.

[Cieslak 79] TH. J. CIESLAK, P. J. HICKSON. Analysis and simulation results of No. 4 ESS network. *IEEE Transactions on Communications,* Vol. COM 27, No. 1, January 1979, pp. 1–9.

[Grinsec 83] GRINSEC. *Electronic Switching* (2 volumes), *Studies in Telecommunications,* Vol. 2, North-Holland, Amsterdam, 1983.

[Inose 81] H. INOSE. *An Introduction to Digital Integrated Communication Systems,* Peter Peregrinus, London, 1981.

[Irland 74] E. A. IRLAND, U. K. STAGG. New developments in suburban and rural ESS. *Proceedings of the International Switching Symposium,* Munich, 1974, Paper 512.

[Joel 79] A. E. JOEL, JR. Digital switching, how it has developed. *IEEE Transactions on Communications,* Vol. COM 27, No. 7, July 1979, Part 1.

[ITT1240 81, 85] ITT 1240, Digital exchange. *Electrical Communication, Special Issue.* Vol. 56, Nos 2–3, 1981; Vol. 59, Nos 1–2, 1985.

[Lotze 79] A. LOTZE, K. ROTHMAIER, R. SCHELLER. TDM versus SDM switching arrays, a comparison. *Proceedings of the 9th International Teletraffic Congress,* Torremolinos, 1979.

[Rothmaier 81] K. ROTHMAIER, R. SCHELLER. Design of economic PCM arrays with a prescribed grade of service. *IEEE Transactions on Communications,* Vol. COM 29, No. 7, July 1981.

[Syski 86] R. SYSKI. *Introduction to Congestion Theory in Telephone Systems, Studies in Telecommunications,* Vol. 4, North-Holland, Amsterdam, 1986 (reprinted from Oliver and Boyd, London, 1960).

Chapter 4

BLOCKING CALCULATIONS IN SWITCHING NETWORKS

In this chapter we shall discuss the classical methods of calculating the probability of internal blocking in connecting networks. These methods enable evaluations to be made successfully under certain assumptions with a certain amount of calculation effort. It is important above all to understand the nature and to grasp the limitations involved in the assumptions made. Specialists will also have to acquire experience and will need ingenuity.

Section 4.1 gives the assumptions which define precisely the class of systems under study and enable the analysis to be simplified.

Section 4.2 develops the Jacobeüs method and Section 4.3 gives the equivalent Lee–Legall systematic procedure method. This involves a procedure which in principle enables the evaluation to be carried out completely mechanically.

The critical factor in these calculations is the choice of the distribution to be used; it is clear at the subscriber stage but is not at all obvious in the center of the network. Section 4.4 provides some help in making this choice.

Other approaches have been attempted; in Section 4.5, the "effective accessibility method" of Lotze's research team is discussed.

The assumptions and the complexity of the calculation in the general case explain why simulation is so widespread; this topic is covered by Section 4.6.

We shall talk of switching networks without being any more specific about them; because of the equivalence mentioned in the preceding chapter, the techniques apply to space-division or time-division networks (in this connection see the notes and conclusions of Section 4.7).

The reader may ask why the birth and death processes, whose efficiency was shown in Section 2.4, are not used. In principle, there is no reason to prevent such an approach; this has in fact been attempted [Elldin 56]. Unfortunately, the numerical treatment of the system of equations involved makes it impossible to envisage the solution of a case of interest in practice. Therefore, approximate methods have to suffice; experience shows that they do not introduce disturbing errors. Another approach, the so-called "thermodynamic" approach will be briefly presented in Chapter 6.

4.1 Calculation assumptions

A first set of assumptions defines the class of studied systems, i.e. the functioning and operation of the network. They do not pose any difficulty; either the network under study belongs to this class or it does not.

Assumption A1. The switching network (the link system) functions according to the principle of end-to-end selection (or "conditional selection"); a call only occupies a path if it is entirely available and, for this, tries all the possible paths. This search (which most often takes place in the memory of the control computer) is assumed to be instantaneous.

We restrict ourselves to pure loss systems; a source which cannot be served receives the busy tone and definitely abandons its attempt. This amounts to assuming that the phenomenon of reattempts which cannot be avoided does not disturb the nature of the traffic.

Networks operated with delay will be discussed in Chapter 6.

Assumption A2. Let us define the concept of a stream of traffic; a homogeneous group of sources offers a traffic stream to a homogeneous group of outputs (a subscriber or a trunk group). Thus, each subscriber offers a different traffic stream to each outgoing trunk group.

If a state of traffic streams is given (j_1 calls in stream $1, \ldots, j_n$ calls in stream n), in general there will exist several ways of positioning the communications in the network. We assume the resulting configurations to be equally probable.

This assumption implies a symmetry in the structure and wiring of the network. This also implies a random choice of paths from those which are offered to each new call (note that this choice may not be the most favorable) (see Chapter 6). This rule is generally obeyed.

Assumptions A3 and A4 define the structure of the traffic offered to the network.

Assumption A3 All streams of calls sharing a common resource (a matrix or a link) originate from identical sources. Depending on the case, Erlang, Engset or Bernoulli traffic will be assumed; this suffices to describe their stochastic behavior. The holding time distribution is no longer considered since the state probabilities are known to be independent of it for loss systems.

The case of sources offering nonidentical traffic will be dealt with in Chapter 5.

Assumption A4 (symmetry and independence of the flows). Each source or group of sources offers a traffic corresponding to one or more traffic streams independent of each other. The random nature of this traffic is not perturbed by the network (i.e. the probability of rejection remains low). In particular, different streams of traffic from the same stage are independent.

These are reasonable assumptions. Some structures will be more stable than others to their nonobservance (for imbalance and correlation between flows, see Chapter 5).

In contrast, the following assumptions are only justified by the simplification which they bring to the calculation; they have to be accepted even if it entails verifying the consequences.

Assumption A5. It is assumed that successive stages of the switching network are independent of each other from the viewpoint of the occupation situations. This assumption cannot hold strictly since conditional selection imposes strong correlations (e.g. a matrix has the same number of occupied inputs and outputs). It should be noted that "dependence" takes two distinct forms: statistical dependence between occupation states and functional dependence between traffic distributions. We shall return to this distinction in Section 4.4.

Assumption A6. Similarly, it will be assumed for the time being (the Lee method) that there is independence between the different links of the same stage. This implies a "Bernoulli traffic" and is no more justified in general than assumption A5.

Finally, from the viewpoint of the evaluation, it is assumed that the network is in statistical equilibrium.

It would then be possible to describe the system (network plus sources) by a set of discrete state variables as in Chapter 2. The size of such a multi-dimensional representation would in general prohibit any benefits being derived from this even using the symmetry relationships in order to reduce the numbers (see for example [Benes 78]). Nevertheless, its potential existence proves (see [Benes 65], Chapter 8) that the evolution of the system follows a steady-state Markov chain with a finite state space. The following consequences can be derived from this.

We assume that the network is observed for a time T. $A(T)$ calls occur of which $R(T)$ are rejected. The limits

$$\alpha = \lim_{T\to\infty} \left\{\frac{A(T)}{T}\right\}$$

and

$$\beta = \lim_{T\to\infty} \left\{\frac{R(T)}{T}\right\}$$

exist (arrival and rejection rates) and so does

$$B = \lim_{T \to \infty} \left\{ \frac{R(T)}{A(T)} \right\} = \frac{\beta}{\alpha} \tag{4.1}$$

B is the blocking probability. Equation (4.1) can be observed by simulation.

Several definitions could be given for internal blocking corresponding to different quantities which can be observed under different conditions. We have already defined the blocking probability or call congestion (eqn (4.1)); we can also introduce the concept of "time congestion" which is denoted as B_T (the proportion of the time during which it is not possible to connect two given ends).

Similarly, point-to-point and point-to-group blocking can be defined (cf. Chapter 3).

Finally, it should be noted that the concept of blocking may or may not involve the occupation of one end (or even both ends).

Thus, blocking seen from an incoming circuit is due to the occupation of the called subscriber or internal blocking if the subscriber is idle. The term "internal blocking" is reserved for the blocking probability between free ends. We denote P as the probability that the called party is busy. The blocking experienced by the incoming circuit will then be

$$B_{\text{total}} = P + (1 - P)B$$

In the following, the internal blocking probabilities will be calculated point to point. It should be noted that because of the assumptions made (A3 and A4) the probability does not depend on the chosen pair of ends. Furthermore, the individual blocking situations experienced are identical for all the pairs connected to the same pairs of matrices; thus the term "blocking between matrices" can be used. In fact, it would be more simple to define traffic between matrices rather than between individual ends.

We denote x as a state of the system and $P(x)$ its probability; $x \in X$, the (finite) set of states. Having chosen a pair of end matrices, let Y be the set of states where they cannot be connected and $\lambda(x)$ the rate of arrivals in state x.

$$B = \frac{\sum_{x \in Y} \lambda(x) P(x)}{\sum_{x \in X} \lambda(x) P(x)} \tag{4.2}$$

and

$$B_T = \sum_{x \in Y} P(x) = P(Y) \tag{4.3}$$

If λ does not depend on x, then B and B_T are identical.

4.2 The Jacobeüs method

We shall explain the method using the example of a three-stage network (Fig. 4.1(a)); we note that this network is equivalent to that in Fig. 4.1(b) where a path is sought to an outgoing direction (route or trunk group). The Jacobeüs method is often expressed in these terms; the equivalence between a point-to-point selection (Fig. 4.1(a)) and the corresponding point-to-group selection (Fig. 4.1(b)) was pointed out in Chapter 3 provided that a single circuit supplies each matrix.

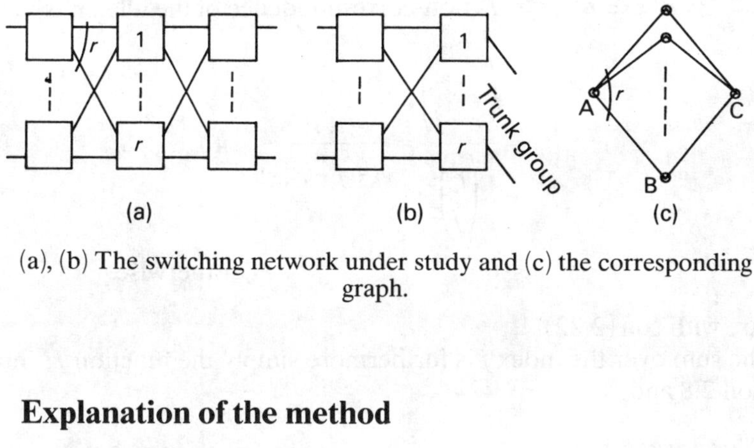

Fig. 4.1 (a), (b) The switching network under study and (c) the corresponding channel graph.

4.2.1 Explanation of the method

Starting from eqn (4.2) or (4.3) we use the channel graph (Fig. 4.1(c)). The network states are fully specified by the list of busy AB and BC links. The set of all the states can be partitioned according to the occupancy of the AB–BC links which the terminal pairs can seize:

$$X = \bigcup_{i,j} X(i, j)$$

where $X(i, j)$ designates the subset of the states such that i (or j) out of the r links AB (or BC) concerned are occupied. The symmetry assumption A4 is made and also the assumption A5 of (statistical) independence between stages. Then,

$$P\{x \in X(i, j)\} = P_1(i)P_2(j)$$

where P_1 and P_2 give the occupancy probabilities of stages AB and BC and are given by the usual teletraffic formulae (functional independence).

If Y designates the set of "blocking" states, we have (using eqn (4.3) which is simpler to handle)

$$B_T = \sum_{x \in Y} P_1(i) P_2(j)$$

and by conditioning the state of occupation (i, j)

$$B_T = \sum_{i,j} P_1(i) P_2(j) P(x \in Y \mid i, j)$$

The last term is the blocking probability knowing that i and j links are occupied. In the example in Fig. 4.1,

$$P(x \in Y \mid i, j) = P(\text{non-correspondence of the idle links})$$

$$= \frac{\binom{i}{r-j}}{\binom{r}{j}} = \frac{i!\, j!}{(i+j-r)!\, r!} \qquad \text{if } i+j > r$$

$$= 0 \qquad \text{otherwise}$$

(compare with eqn (2.22)).

The sum over the index j is furthermore simply the function H introduced in Section 2.8 and

$$B_T = \sum_{i \leqslant r} P_1(i) H(r-i)$$

$$= \sum_{i=1}^{r-1} P(r-i\ AB \text{ links idle, } i \text{ busy}) P(\text{corresponding } r-i\ BC \text{ links busy}) \quad (4.4)$$

This formula can be immediately understood intuitively and is often used to write down the blocking directly by this method. The approach adopted here nevertheless allows a more general formulation which can be extended to other structures. In the general case, the use of eqn (4.2) for writing the call congestion leads to cumbersome calculations; it is preferable to express it by noting that it is the time congestion due to adjacent sources of the selected ends because of the Poisson nature of the traffic.

We give a few examples of the application; the choice of the distribution to be used for P_1 and P_2 remains the crucial point.

We designate A and C as the traffic offered to the matrices A and C and a and c as the link occupancies; $= A/r$ (α designates the Engset distribution parameter). The expressions for P_i and H are those in Chapter 2, eqns (2.12), (2.17), (2.23) and (2.24).

4.2.2 Some examples (three-stage network)

Bernoulli traffic at *A* and Erlang traffic at *C*

Each matrix of stage A receives n independent sources, subscribers or circuits of different trunk groups (each trunk group is split over the matrices so that a single circuit of each trunk group supplies one matrix and there is no sharing between inputs). The independence of the sources justifies the Bernoulli distribution:

$$P_1(i) = \binom{n-1}{i} a^i (1-a)^{n-1-i} \qquad n = r$$

It should be noted that $n-1$ and not n is used; the input for which a path is sought is obviously free. If the first stage concentrates the traffic, the equation no longer applies; it is necessary to use the Engset distribution. If it is being expanded $(n<r)$, a will be replaced by the traffic per source $e(a = en/r)$ and $P(i) = 0$ if $n < i \leqslant r$.

The BC stage receives the traffic from a large number of sources which share its links; the Erlang distribution is then used (in an approximate manner). H is given by eqn (2.33):

$$B = \sum_i \binom{m}{i} a^i (1-a)^{m-i} E(r-1, C) \frac{\sum_k (C^k/k!)}{C^i/i!}$$

We have set $m = n-1$ in order to simplify the notation. $E(N, A)$ naturally designates the Erlang loss (N circuits and offered traffic A).

The reader is left to simplify this expression by astute change of subscripts after expansion of the term $1-a$. We finally find that

$$B = \frac{E(r, C)}{E(n-1, C/A)} \tag{4.5}$$

According to above, eqn (4.5) gives the percentage of time during which the neighboring $n-1$ sources occupy all the usable paths; this is call congestion, the total occupation probability at the random time of arrival of a call emanating from an idle source. Depending on whether a path is being sought between two points or between a point and a trunk group (covering the loss in the trunk group), the correction $(n, r \rightarrow n-1, r-1)$ will be made or not.

If the first stage concentrates traffic, the Bernoulli distribution is replaced by the Engset distribution and

$$B = \sum_{j=r}^{n} \frac{\binom{r}{j} \alpha^j}{\sum_{i \leqslant r} \binom{r}{i} \alpha^i} \frac{E(r, C)}{E(j, C)} \tag{4.6}$$

It should be noted that, because of the concentration r links may be engaged. In practice, there will be a small error in this type of expression when eqn (4.6) is replaced by eqn (4.5). Some researchers suggest using the Bernoulli distribution for P_1, correcting $P_1(r)$ in order to allow for the possibility of blocking the trunk group AB:

$$P_1(r) = \binom{n}{j} a^j (1-a)^{n-j}$$

If the concentration is very high, the Erlang distribution can be used for P_1.

Bernoulli traffic at both stages

If the independence assumption is justified for the BC stage (e.g. the same configuration as the stage A), then

$$H(j) = c^j$$

For a network with no expansion or concentration ($n = r$),

$$B = \sum \binom{r-1}{j} a^j (1-a)^{r-1-j} c^{r-1-j}$$

$$= \{a + (1-a)c\}^{r-1} \tag{4.7}$$

If the input matrices are undergoing expansion ($n > r$), we have

$$B_T = \{e + (1-e)c\}^{n-1} c^{r-n+1} \tag{4.8}$$

where

$$e = \frac{ra}{n}$$

In fact, it is the traffic per source which is involved; the probability that i AB links are busy is given by the probability that i sources out of $n-1$ (and not n, since it is known that the source for which the blocking is sought is idle) are busy.

It does not seem necessary to give any further application examples. We hope that we have made the application method sufficiently clear to the reader that he can use it himself. In [Syski 86] and [Bear 76] a large number of applications are given.

Much effort has been spent in extending the field of application of this method; complex networks with more than two link stages taking into account gradings etc. have been considered. Bearing in mind the disappearance of this type of wiring, we shall dispense with this discussion (see above-mentioned references). We give an example of calculation on a nonseries–parallel network with four matrix stages which shows the generalization of the method.

4.2.3 A four-stage network

In general, the Jacobeüs method consists of decomposing the graph using the assumption A5 in order to write the state probabilities (Fig. 4.2):

$$P(\text{no path}) = \sum_j P(j \text{ idle links})\, P(\text{no path within } B \mid j)$$

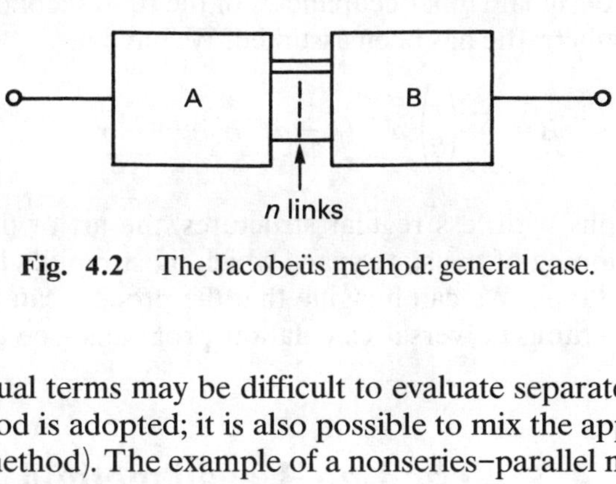

Fig. 4.2 The Jacobeüs method: general case.

The individual terms may be difficult to evaluate separately and therefore an iterative method is adopted; it is also possible to mix the approaches (in particular, the Lee method). The example of a nonseries–parallel network (Fig. 4.3) will serve as an illustration.

The blocking probability is evaluated by conditioning the state of the outputs of the first matrix. Each idle AB link gives access to a matrix B originating from a channel graph with three stages. If j AB links are idle, a given matrix C

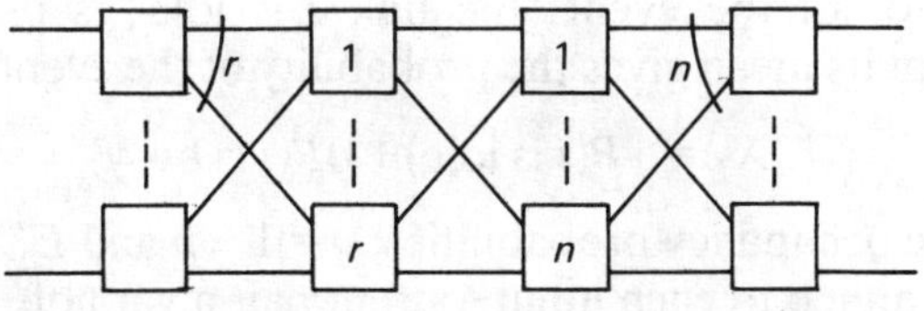

Fig. 4.3 A nonseries–parallel network with four stages.

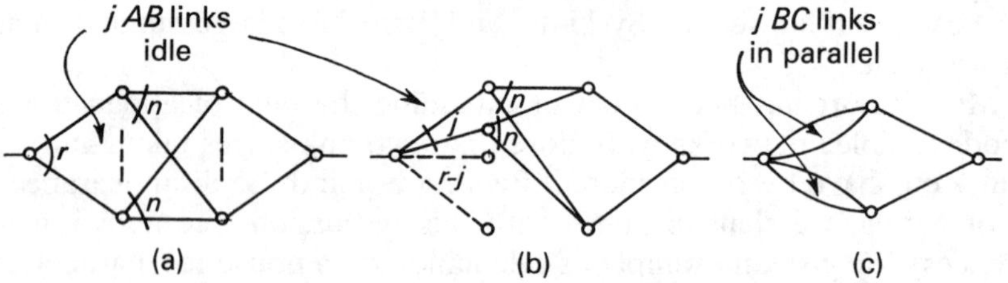

Fig. 4.4 Conditional channel graph.

may be reached by the j BC links in parallel (Fig. 4.4(b)). The channel graph (conditioned by the event "j AB links idle") takes the form of Fig. 4.4(c).

The Lee method (see the next section) immediately gives the following result which could have been derived by the Jacobeüs method (this example shows the flexibility in combining the two approaches:

$$P(\text{no path} \mid j \text{ idle links}) = \{b^j + (1 - b^j)c\}^n$$

where a, b and c designate link occupancies of the first, second and third stages. Of course, Bernoulli traffic has been assumed. We have

$$B = \sum_{j=0}^{r} \binom{r}{j} a^{r-j}(1-a)^j\{b^j + (1-b^j)c\}^r$$

For complex graphs with less regular structures, the art of the method lies in finding the partitioning of the state space which most rapidly leads to graphs of the series–parallel type. We can imagine that the process can be automated for large graphs by writing universal calculation programs (see [Kodaira 76] and [Villar 72]).

4.3 The Lee–Legall method

This calculation method, which is equivalent to the Jacobeüs method, gives a systematic procedure for writing down the blocking probabilities.

We attach to each link i of the network a random variable X_i which takes the value 1 if the link is idle and otherwise takes the value 0. In probability theory the indicator of the event "the link i is idle" is used. The indicator function is such that its mean gives the probability of the event:

$$E(X_i) = 1P(i \text{ is idle}) + 0P(i \text{ is busy}) \tag{4.9}$$

We denote by p the occupancy probabilities; $q = 1 - p$ and $E(X_i) = q_i$.

Similarly, we attach to each input–output pair a variable Z, the indicator of the event "one idle path at least between the ends". Because of symmetry assumptions, the distribution of Z does not depend on the selected pair.

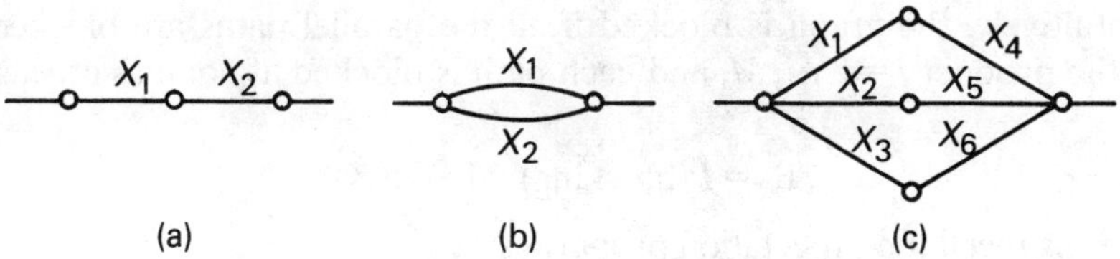

Fig. 4.5 Elementary graphs: (a) series; (b) parallel; (c) series–parallel.

Similarly, p and q are assumed to depend only on the stage (identical for all the links of the stage) and are indexed by the rank of the stage.

The value of Z will be deduced from X_i. We start with the elementary cases of the channel graphs in Fig. 4.5.

For the graph in Fig. 4.5(a), we write

$$Z = X_1 X_2$$

since the path is idle only if each link is idle.

For the parallel graph (Fig. 4.5(b)), $1 - Z$ is evaluated more easily; it is equal to 1 if no path exists:

$$1 - Z = (1 - X_1)(1 - X_2)$$

Combining the two results, the indicator of the graph in Fig. 4.5(c) is

$$1 - Z = (1 - X_1 X_4)(1 - X_2 X_5)(1 - X_3 X_6)$$

We are now in the position to write down the indicator function for any simple series–parallel graph (by a simple graph we mean the graph of a series–parallel network) (Fig. 4.6).

$$Z = 1 - \prod_{j=1}^{r} (1 - Y_j)$$

where Y_j designates the variable attached to the jth path; $Y_j = X_{1j} X_{2j} \ldots X_{Mj}$ and

$$Z = 1 - \prod_{j=1}^{r} \left(1 - \prod_{i=1}^{M} X_{ij}\right) \tag{4.10}$$

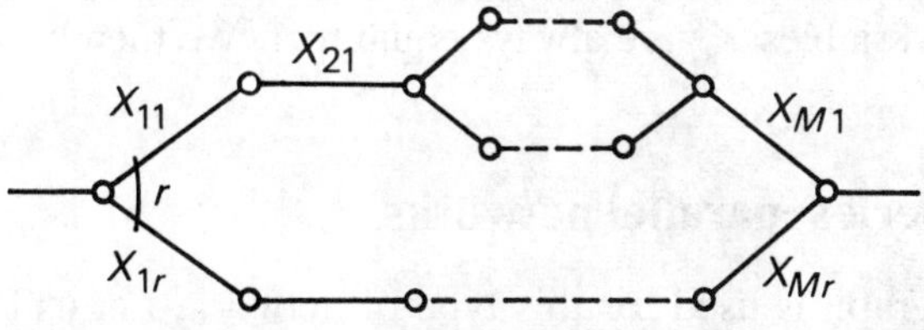

Fig. 4.6 Any series–parallel graph.

Intuitively, the graph is blocked if all the parallel paths are blocked and hence the product $j=1,\ldots,M$, and each path is blocked as soon as at least one arc is.

$$B_T = P(\text{blocking}) = 1 - E(Z)$$

Let us recall the expectation properties:

$$E(A+B)=E(A)+E(B) \tag{4.11}$$

$$E(AB)=E(A)E(B) \text{ if } A \text{ and } B \text{ are independent}$$

Here,

$$B = 1 - E\left\{1 - \prod_j (1-Y_j)\right\}$$

$$= E\left\{\prod_j (1-Y_j)\right\}$$

We now impose assumption A5 of the independence between stages and temporarily add to it assumption A6 of the independence between links of the same stage. Then, applying eqn (4.11), we have

$$B = E\left\{\prod_j (1-Y_j)\right\} = \prod_j \{1 - E(Y_j)\}$$

$$= \prod_j \left\{1 - \prod_i E(X_{ij})\right\} \tag{4.12}$$

so that finally

$$B = \left(1 - \prod_{i=1}^{M} q_i\right)^r \tag{4.13}$$

For example, the network in Fig. 4.1 will have a blocking probability given

$$B = (1 - q_1 q_2)^r = \{a + (1-a)c\}^r$$

This is eqn (4.7) since assumption A6 is the same as considering Bernoulli traffic.

If there is expansion ($n < r$), at most n links out of r links of stage 1 may be busy and $r-n$ variables X_{1j} are always equal to 1. We then have eqn (4.8) once more.

4.3.1 Nonseries–parallel networks

The canonical graph is used for this type of network. Let us isolate two paths using the same link AB (Fig. 4.7, which shows part of the graph of the switching network of Fig. 4.3).

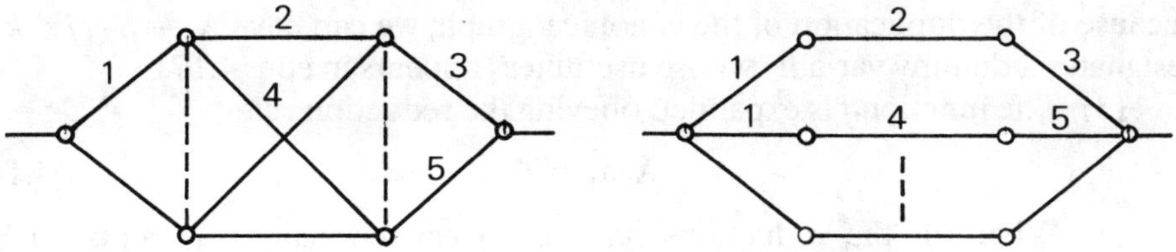

Fig. 4.7 Graph of a nonseries–parallel network.

The indicator function Z contains terms relating to the two paths

$$Y_1 = X_1 X_2 X_3$$
$$Y_2 = X_1 X_4 X_5$$

Y_1 and Y_2 are equal to 1 if the paths are idle and Z contains the composite term giving the availability of at least one of the two:

$$(1 - Y_1)(1 - Y_2) = 1 - Y_1 - Y_2 + Y_1 Y_2$$

The term $Y_1 Y_2$ relates to the availability of the two paths at the same time which occurs if link 1 is idle ($X_1 = 1$):

$$Y_1 Y_2 = X_1^2 X_2 X_3 X_4 X_5$$

It is thus necessary to use the rule $X_1^2 = X_1$ since we are dealing with indicator functions.

To the changes which convert eqn (4.10) to eqn (4.12), we therefore add the rule

$$\left.\begin{aligned} E(X_i X_j) &= EX_i EX_j = q_i q_j \\ &= EX_i \quad\;\; = q_i \end{aligned}\right\} \quad \begin{aligned} &\text{if } i \neq j \\ &\text{if } i = j \end{aligned}$$

4.3.2 General statement

(1) We operate on the canonical graph which only differs from the normal channel graph for nonseries–parallel networks. Each link has a variable X attached to it whose expectation is the probability of finding it free.

(2) We define the "basic function" as

$$f(s) = 1 - \prod_{j=1}^{N} (1 - Y_j s) \tag{4.14}$$

where j designates the index of the paths (N paths in total) and Y_j the variable relative to the jth path:

$$Y_j = X_{1j} X_{2j} \ldots X_{Mj}$$

Because of the duplication of the canonical graph, we can have $X_{ij} = X_{ik}$, $j \neq k$. s designates a dummy variable whose usefulness appears in eqn (4.17).

(3) The function f is expanded obeying the reduction rule

$$X_i X_i = X_i \tag{4.15}$$

(4) When all the reductions have been carried out, X_i is replaced by $q_i = 1 - p_i$ (where p_i is the occupancy probability, i.e. the traffic carrried by the links). The function obtained in this way, which is denoted $F(s)$, is a polynomial in s of degree N, the maximum number of paths which may exist.

(5) The probability of internal blocking is given by

$$B = 1 - F(1) \tag{4.16}$$

The probability that there are exactly k paths is

$$P(\text{exactly } k \text{ paths}) = \frac{(-1)^{k+1}}{k!} F^{(k)}(1) \tag{4.17}$$

and the mean number of available paths is $F'(0)$.

Proof

The construction of the function f (and therefore F) made in the preceding section justifies the statement intuitively. More precisely, the function F can be expanded:

$$F(s) = \sum_{j=0}^{N} (-1)^{j+1} S_j s^j \tag{4.18}$$

where S_j is the sum of the probabilities of exactly j idle paths.

A classical lemma due to Poincaré [Feller 71] makes it possible to write directly

$$P\left(\bigcup_{i=1}^{N} y_i\right) = \sum_{j=1}^{N} (-1)^{j+1} S_j$$

where y_i is the event "the branch i is idle".

Equation (4.16) can be deduced from this immediately. Equation (4.17) can be deduced from eqn (4.18) by deriving

$$F^{(n)}(s) = \sum_{j \geqslant n} (-1)^{j+1} s^{j-n} S_j \frac{j!}{(j-n)!}$$

and therefore

$$\frac{(-1)^n}{n!} F^{(n)}(1) = \sum_{j \geqslant n} (-1)^{j+n+1} \binom{j}{n} S_j$$

This is the probability of finding exactly n events out of the y_i (see [Feller 71], Chapter 2).

We should point out that, as for the Jacobeüs method, it is possible to automate and computerize the method; this is what NEASIM [Grantges 64] does, a program which has dimensioned the systems of the ESS range [ESS1 79], [ESS4 79].

Finally, calculations which seem to be inextricable may be largely simplified by using transformation rules (see [Legall 57]).

4.3.3 Allowing for dependences (in the stage)

We show how it is possible to discard the assumption A6 of independence between links of the same stage. The following theorem is due to Legall.

Theorem. In order to obtain the blocking probability in a network (time congestion), the probability obtained by the Lee method (eqn (4.16)) is expanded. In the resulting polynomial the terms $p_i^k q_i^j$ (where i designates the stage) is replaced by the expression giving the probability of detecting k given links busy and j given links idle. We then obtain the "exact" expression (i.e. without assumption A6) for the blocking probability.

For example, starting from eqn (4.7) which can be derived from the method under discussion, we write

$$B = \{p_1 + (1 - p_1)p_2\}^r$$

$$= \sum_{j=0}^{r} \binom{r}{j} p_1^{\,j} \{(1 - p_1)p_2\}^{r-j}$$

We assume an Erlang distribution at the second stage; we replace p_2 by

$$p_2^{\,k} \to \frac{E(r, C)}{E(r-k, C)}$$

where C designates the traffic offered to the matrix C. Then,

$$B = E(r, C) \sum_{j=0}^{r} \binom{r}{j} p_1^{\,j} (1 - p_1)^{r-j} \frac{j!}{C^j} \sum_{k=0}^{j} \frac{C^k}{k!}$$

We note that

$$\sum_{j=0}^{r} \sum_{k=0}^{j} = \sum_{i=j-k=0}^{r} \sum_{j=i}^{r}$$

This is the transformation which in Section 4.2.2 made it possible to derive eqn (4.5). The two methods actually give the same result!

Proof

(1) We consider an isolated trunk group of r circuits. We denote $|p^i q^j|$ as the probability of finding i defined circuits busy and j circuits idle. $P(n)$ gives the probability of occupation of any n circuits. The proof of eqn (2.22) is easily generalized and

$$|p^i q^j| = \sum_{n=i}^{r-j} P(n) \frac{\binom{r-i-j}{n-i}}{\binom{r}{n}} \tag{4.19}$$

Let $|\boldsymbol{pq}|_j$ be the vector of elements $|p^i q^j|$, $i=0, \ldots, r$, $|p|$ be the vector of elements $|p^i|$ (i.e. $j=0$), $i=0,\ldots,r$, $\boldsymbol{P}$ be the vector of elements $P(n)$, $n=0,\ldots,r$, and $\mathbf{A}_j$ be the square matrix $(r+1)\times(r+1)$ with elements

$$a_{i,k} = \frac{\binom{r-i-j}{k-i}}{\binom{r}{k}} \quad \text{for } 0 \leqslant i \leqslant r, \ i \leqslant k \leqslant r-j$$

$$= 0 \quad \text{otherwise.}$$

Equation (4.19) takes the matrix form

$$|\boldsymbol{pq}|_j = \mathbf{A}_j \boldsymbol{P}$$

$$|\boldsymbol{p}| = \mathbf{A}_0 \boldsymbol{P}$$

The matrix $\mathbf{A}_j$ is triangular and therefore can be inverted:

$$|\boldsymbol{pq}|_j = \mathbf{A}_j \mathbf{A}_0^{-1} |\boldsymbol{p}| \tag{4.20}$$

All the combinatorial properties of the system have been concentrated in the matrix, the random properties being in $|\boldsymbol{p}|$.

(2) Any occupation state S of the trunk group may be described in terms of unions or intersections of elementary events of the type "j given circuits busy, the other $r-j$ being idle". We therefore have

$$P(S) = \sum_{i=0}^{r} |p^i q^{r-1}| d_i = \sum_{i=0}^{r} |p^i| c_i \tag{4.21}$$

According to eqn (4.20), d_i and c_i are constants, independent of the nature of the traffic. In particular, they can be obtained by assuming a Bernoulli distribution and by expanding $P(S)$ in this case:

$$P(S, \text{Bernoulli case}) = \sum_i c_i p^i$$

The c_i can be derived by identification.

(3) The state of a switching network depends on the state of one or more groups of circuits. If interest lies in a particular trunk group (links between the first and second stage), the same statement can be made; any state, and in particular the "blocking" state, can be expressed as a combination of elementary states of the trunk group considered and the probability of a state can be written as a linear combination of the probabilities of the elementary states (eqn (4.21)):

$$B = \sum_i b_i |p^i|$$

where the b_i, independent of the nature of the traffic, are obtained by assuming Bernoulli traffic on this trunk group, i.e. by using the Lee method since then $|p^i| = p^i$. The Legall rule consists of identification in eqn (4.21) by replacing p^i or $p^i q^j$ by the exact expression $|p^i q^j|$. In general, B depends on several trunk groups (links of the first stage, second stage etc.); eqn (4.21) is applied iteratively.

Note that the following corollary can be proved [Legall 57]; any inequality concerning the network structures (with the same number of stages and links equally loaded) derived under the Bernoulli assumption immediately extends to any type of traffic.

This result is a consequence of eqn (4.21). It makes it possible to compare structures with modest mathematical techniques. Care should be taken in extending it to the comparison of dissimilar structures.

4.4 Dependence between stages; the choice of distribution

The above-described methods all more or less neglect the obvious dependence between the stages of the network. The dependence assumes two different forms.

- It modifies the distribution of the occupation (this is called functional dependence), the mixing of the traffic streams and rejection making it difficult to predict the distributions to be used at each stage.

- It introduces a dependence between the occupation states, the most obvious manifestation being that it is impossible to have a different number of busy inputs and outputs. This is called statistical dependence.

The latter acts favorably, causing the calculated blocking to be higher than its true value.

"Functional" dependence

The Jacobeüs and Lee–Legall methods require the choice of a probability distribution for the occupation of the links of a stage. Examples of this choice have been given in Section 4.2.2 for elementary cases.

The concept of a "traffic group" is fundamental from this viewpoint; we have already defined the concept of "traffic stream"; the links of a stage which can be shared by a traffic stream form what is called a traffic group. Thus, in the network in Fig. 4.3, all the connections between a pair of matrices *A* and *D* are shared by all the links of stages *BC* which therefore form a traffic group.

In contrast, a traffic group is in general shared by several streams; all the pairs of matrices *A*–*D* take the same connections from the stage *BC*.

The problem is therefore to determine the vector $|\boldsymbol{pq}|$ or, from eqn (4.20), $|\boldsymbol{p}|$ for a traffic group [Legall 57].

Let us assume that the traffic group considered (situated at the first stage, say) carries a mixture of traffic streams so that in this group all the occupation situations are possible. Then, except for the perturbation introduced by the network (this perturbation is the blocking possibility which is small), the probability of detecting that j links are busy, i.e. $|p^j|$ is the same as for an isolated trunk group accepting traffic of the same kind; this will be the Erlang, Engset, Bernoulli or other distribution.

This is the concept of "the most concentrated stage". If it does not exist, a virtual stage is added to the network to perform its function. The value of this stage is that it is possible to calculate its occupation probabilities.

Let us now assume that this stage is adjacent to a stage in expansion (the second stage). In the latter, all the occupation situations are not possible. Nevertheless, there are as many busy inputs as outputs in the matrices common to the two stages. Then,

$$|p_2{}^j| = \left(\frac{L_1}{L_2}\right)^j |p_1{}^j|$$

(the notation is explained in Fig. 4.8).

This relationship may be proved directly. More rapidly, it can be derived from Section 4.3.4; this is obviously true for Bernoulli traffic (the multiplication factor is the ratio of the carried traffic) and the arguments in this section extend to any kind of traffic.

The distribution to be chosen for the most concentrated stage depends on the sources, the traffic, the concentration etc. The preceding argument enables

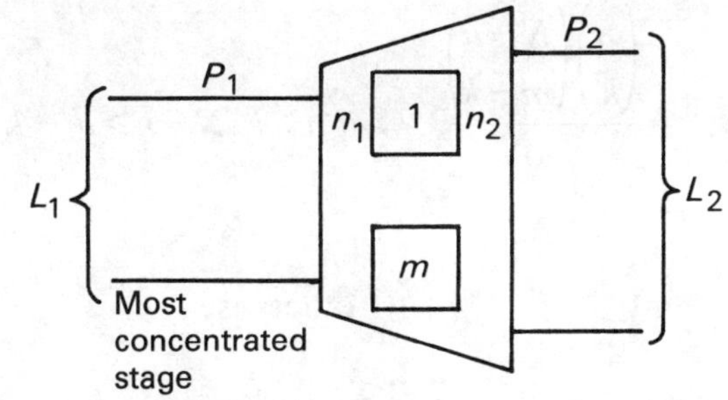

Fig. 4.8 The "most concentrated" stage and its neighbors.

$$L_1 = mn_1$$
$$L_2 = mn_2 > L_1$$
$$p_2 = \frac{p_1 L_1}{L_2}$$

the distributions to be used in the adjacent stages and step by step over the entire network to be derived from the distribution for the concentrated stage.

Perhaps, it is not very statisfactory to have chosen the distribution over a stage "arbitrarily". In fact, only the distribution over the traffic sources is given. It should be possible to derive the traffic in the network from the distribution over the traffic sources. The following approach enables this calculation to be carried out effectively, although it does introduce complications [Bininda 67].

Let us assume a series–parallel channel graph with M stages (Fig. 4.9).

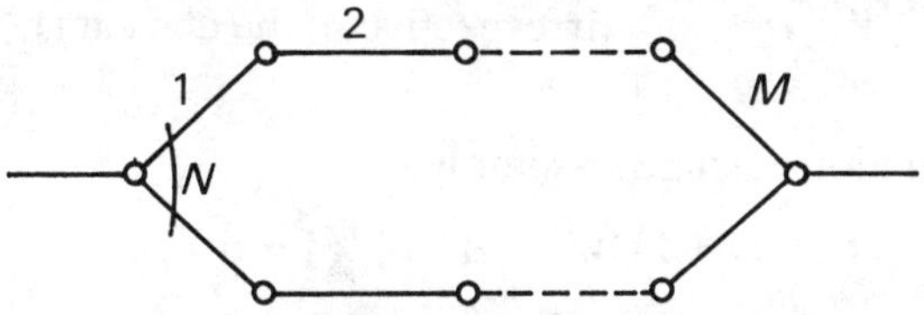

Fig. 4.9 The Bininda algorithm.

We shall call $w_m^{(i)}$ the probability that m paths exist out of the N possible paths up to the stage i (scanning is from left to right by stage), $W_n^{(i)}$ the probability that n out of the N links are busy and $g_{k,m,n}$ the conditional probability that k paths are idle after the stage i, given that m are idle up to then and that n links are idle at the stage. $g_{k,m,n}$ can be evaluated from eqn (2.22):

$$g_{k,m,n} = \frac{\binom{n}{k}\binom{N-n}{m-k}}{\binom{N}{m}} \qquad \text{if } n \geqslant k,\ N-n \geqslant m-k$$

$$= 0 \qquad \text{otherwise}$$

Then,

$$w_k^{(i)} = \sum_{m=k}^{N} w_m^{(i-1)} \sum_{n=k}^{N-m+k} W_{N-n}^{(i)} g_{k,m,n}$$

$$B = w_0^{(m)} \tag{4.22}$$

This equation is used directly if the $W^{(i)}$ are known. Here, we introduce the function ϕ; $\phi_k^{(i)}$ is the probability that k paths are idle seen from the stage i (i.e. taking into account stages $1, 2, \ldots, i-1, i+1, \ldots, M$). It is related to $W^{(i)}$. In the example in Fig. 4.9, $M = 4$:

$$\phi_k^{(3)} = \sum_{j,m,n} W_{N-j}^{(1)} g_{p,j,n} W_{N-n}^{(2)} g_{k,p,m} W_{N-m}^{(4)} \tag{4.23}$$

If j links are busy, the blocking probability seen from the stage i is

$$\sigma_j^{(i)} = \sum_n g_{0,N-j,n} \phi_n^{(i)} \tag{4.24}$$

and

$$B = \sum_k W_k^{(i)} \sigma_k^{(i)} \qquad \text{(irrespective of the chosen } i\text{)} \tag{4.25}$$

The evolution equation at stage i is simply

$$(j+1) W_j^{(i)} = A^{(i)} W_j^{(i)} (1 - \sigma_j^{(i)}) \tag{4.26}$$

(Poisson arrivals have a rate $A^{(i)}$ and communication durations are exponential).

The method will therefore be iterative.

(1) We start from an arbitrary distribution for the $W^{(i)}$ (Erlang distribution).

(2) Using eqn (4.23), evaluate the $\phi^{(i)}$, $i = 1, \ldots, M$.

(3) Using eqn (4.24), we derive $\phi^{(i)}$ from this.

(4) Equation (4.26) then gives a new distribution for W.

(5) We iterate by returning to (2) until the $W^{(i)}$ no longer evolve. Equation (4.25) applied to any stage then gives the blocking.

As Bininda and Daisenberger [Bininda 67] note, in the general case the convergence of the algorithm cannot be formally proved. However, actual experience shows rapid convergence (a few iterations).

It can be seen that the use of intuition in the choice of distribution considerably reduces the amount of calculation. In practice, all the reported experience and comparisons with simulation show that a judicious choice of distributions is sufficient to give correct results; there is little to gain in precision by using an iterative method as described here.

"Statistical" dependence

In the unanimous view of teletraffic "experts", the assumption of independence is of little importance; it often results in a small error with negligible impact on the dimensioning. It is still more difficult to allow for the statistical dependence, i.e. the correlation between the states of adjacent stages in the general case. We refer the reader to [Legall 57] and [Pippinger 78] which give some idea of the procedures which the rejection of assumption A5 entails. The reader can also prove the effectiveness of the approximations by comparing with simulations for example.

4.5 The method of effective accessibility

This is another approximate method, which is rather unusual, developed in particular by Lotze and his collaborators. It derives its name by analogy with gradings which inspired it.

4.5.1 Gradings

Let there be g groups of k sources to be connected to n circuits ($k < n$). The sources are matrices in a switching network and, in the general case (for any n, gk), it will be a question of homogenizing the traffic, in addition to giving maximum access to each group. Figure 4.10 illustrates the following cases: full availability ($n = \mathrm{k}$); separate groups ($n = gk$); the general case (limited availability). The parameter k is the availability. The interconnection coefficient $H = gk/n$ is the average number of blocks which can be seen by each circuit.

This type of arrangement, which is very widespread in rotary and crossbar systems, is tending to disappear. Their design and equally their evaluation have turned out to be difficult and obscure. A large number of formulas have been suggested for this.

Thus, if the blocking remains low and on the assumption of random choice of circuit, it can be assumed that the traffic distribution on the trunk group obeys the Erlang distribution (A is the offered traffic). Each block has access to k par-

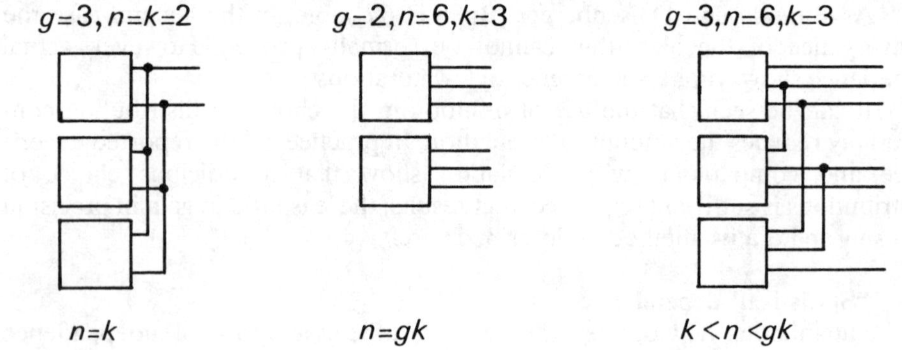

Fig. 4.10 Gradings.

ticular circuits; its probability of failure is then given by eqn (2.22) which is the Palm–Jacobeüs formula:

$$B = \frac{E(n, A)}{E(n-k, A)} \tag{4.27}$$

This formula overestimates the blocking; it has been suggested that it can be improved by replacing A by the value A' which when offered to a trunk group of size n with full availability would carry the same traffic Y as the grading:

$$Y = A'\{1 - E(n, A')\} = A\left\{1 - \frac{E(n, A')}{E(n-k, A')}\right\}$$

This is the modified Palm–Jacobeüs formula which has been well verified.

4.5.2 Principle of the method

We operate on the network in Fig. 4.1, searching for a path to matrix C. Let us call Y_2 the traffic it carries. On average, Y_2 BC links will be busy there and $r - Y_2$ will be idle. We define the effective availability (or accessibility) of the matrix A as the average number of available BC links, in this case $r - Y_2$. Blocking occurs if the corresponding AB links are busy; we use eqn (4.27):

$$B = \frac{E(r, A')}{E\{r - Y_2), A'\}} \tag{4.28}$$

$$A' \text{ such that } Y_1 = A'\{1 - E(r, A')\}$$

Y_1 is the traffic carried by the AB links. In general, Y_2 is not an integer; we shall use the nonintegral version of the Erlang formula (see Appendix).

During successive refinements, Lotze and his collaborators have improved the accuracy of the method. Although it cannot be said that it has a solid theoretical basis, a certain amount of success must be acknowledged (cf. the simulation reported in [Bazlen 74] for example).

In fact, these successes have only been obtained by adjustments in the calculation of effective accessibility and thus can only be justified by the result. It is therefore not certain whether or not a new type of network might require a further correction term!

In the general case, a distinction is made between the two approaches depending on whether a "point-to-point" or "point-to-group" loss probability is required.

4.5.3 Point-to-group selection

The entire network is replaced by a single stage which connects the source to the requested trunk group with the effective accessibility k_{eff}. As above, the accessibility represents the number of circuits which the source may reach (on average). It then remains to apply eqn (4.28).

The accessibility is calculated as follows. Let $i = 1, \ldots, s$ be the successive stages (s denotes the output stage), k_i the number of outputs of stage i matrices, y_i the traffic which they carry and g_i their number.

$$FF(i) = \max\left\{g_i, \prod_{j=1}^{i}(k_j - y_j)\right\}$$

$FF = FF(s)$, the free fan, denotes the number of matrices accessible to the last stage.

$$BF = \max\left(g_s, \prod_{j=1}^{s-1} k_j\right) - FF$$

The busy fan denotes the number of busy output matrices and

$$b = BF\frac{Y}{n}$$

represents the proportion of busy trunk groups demanded (of size n, carrying Y) which cannot be reached. Of these circuits, when a fraction f become idle, and because of idle paths in the network, this may open paths to the FF circuits.

Bazlen *et al.* suggests the following expression for f:

$$f = \frac{1}{k_s} \qquad \text{for a non-linked structure}$$

$$= \frac{g_s i_s - Y_{tot}}{g_s k_s} \qquad \text{otherwise}$$

(Y_{tot} is the traffic carried in the network.) The justification for these formulas is tenuous and (according to Bazlen *et al.*) "their fit had to be checked and confirmed on the basis of many thousands of artificial traffic trials ..." [Bazlen 74].

We thus have

$$k_{eff} = FF + fb$$

(it has been assumed here that a single trunk group circuit supplied each matrix).

4.5.4 Point-to-point selection

The "PPL method" [Lotze 77] takes into account the end stages (Fig. 4.11).

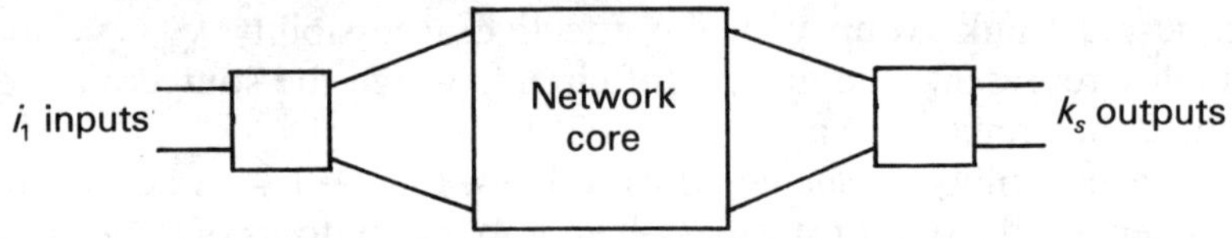

Fig. 4.11 The PPL method.

If x_1 (or x_s) out of the i_1 (or k_s) links are busy, $c_s\{x_s, k_{eff}(x_1)\}$ denotes the point-to-point blocking probability:

$$c_s\{x_s, k_{eff}(x_1)\} = \frac{\binom{x_s}{k_{eff}(x_1)}}{\binom{i_s}{k_{eff}(x_1)}}$$

The proof is the same as for eqn (2.22).

The occupancy probabilities for x_1 and x_s are given by the usual formulas and

$$B = \sum_{x_1=0}^{i_1-1} \frac{P(x_1)}{1-P(i)} \sum_{x_s=0}^{k_s-1} \frac{P(x_s)}{1-P(k)} c\{x_s, k_{eff}(x_1)\}$$

The effective accessibility is calculated as in the preceding section. This is the sum

$$k_{eff} = U + V + W + \ldots$$

where the term U, corresponding to FF, is given by

$$U=(k_1-x_1)\left(1-\frac{Y_2}{k_2}\right)$$

(which can be easily interpreted), the term V is the proportion of the blocked x_1 links which can be reused on their liberation and is given by

$$V=\frac{x_1}{g_s}$$

the term W is the fraction of the second-stage outputs which are busy but can be used as soon as they become idle (and therefore accessible in the sense of the method) and is given by

$$W+\frac{(k_1-x_1)\,y_2}{g_s}$$

etc.

4.5.5 Comments and applications

The method always leads to very simple evaluations, even for large networks.

Nevertheless, it is difficult to find a satisfactory theoretical basis for it. The development of the formulas giving k_{eff} results from empirical considerations which reproduce, without proper explanation, the simulation results.

However, the weakness of the empirical treatment makes it risky to apply the method to a structure unless it has been verified by simulation. It should be said that the method gives very manageable formulas with surprising accuracy in actual cases, showing that it does contain the salient features of the blocking phenomena. The manageability of the formulas has enabled the method to be used in optimization studies (the search of structures which for a given size and blocking probability minimize the number of crosspoints).

Optimization is based on the concept of "transparency" related to accessibility. This is the average number of idle paths in the networks:

$$T=\prod_{j=1}^{s-1} k_s(k_j-y_j)$$

This concept also enables networks to be studied by constructing "reduced models". A network is constructed which is smaller than the network to be studied containing the same number of stages and obeying (the asterisks refer to the reduced model)

$$\frac{T^*}{k_s^* g_s^*} = \frac{T}{k_s g_s}$$

Examples of applications can be found in the references.

4.6 The use of simulation

The above discussion has shown the benefits but also the weaknesses of using probability theory for estimating internal blocking.

- Analytical complexity of almost all the methods; we note in addition that questions concerning the calculation algorithms and their numerical behavior (manipulation of series) have not been mentioned.
- Difficulty in application, e.g. the choice of distribution.
- Validity of the independence assumption for unknown structures.
- Validity of the other assumptions (the symmetry and homogeneity of the sources and the effect of the path-searching algorithms) which it might be desirable to dispense with.
- Finally, calculation readily gives a point-to-point and even a point-to-group loss probability, but the user experiences a more complex grade of service which is difficult to consider analytically (for the sequence of operations requires several path selections connected to each other, see Chapter 3).

All these explain the interest and the frequent use of simulation in the study of blocking in networks.

Simulation here refers to "discrete-event simulation"; the simulation program reproduces the structure of the network, the links and the interconnections.

The behavior of the users is obtained by using random number generators. In this model, it is easy to make measurements as on an actual machine. However, it should not be forgotten that this is a statistical experiment, i.e. the implementation of a complex random process.

The evolution is described with the help of the concept of the event, e.g. a subscriber unhooking, end of dialing and transfer to conversation, and hanging up (the exact list depends on the structure and the operating mode).

Formally, an event is a change in state of a system. The state is described by a set of discrete or continuous variables. Nothing happens between two successive events and the simulation jumps from one point of the time axis to another. Each event may generate other events. Thus the beginning of communication eventually generates a hanging-up.

The simulation monitor connects the events, which it places in the "list of future events" (also called the "pending event set"). This is filled and emptied as the network evolves.

The simulation technique is not confined to studies of switching networks. However, its use for a specific purpose produces particular problems which are worth identifying.

More precisely, the simulator will have to cope with the following problems.

- The large number of future events (conversations in progress at the same time) which produces a large event set whose management takes up an appreciable amount of time.
- The large number of traffic sources (subscribers and circuits); the correct method would most frequently be to simulate each source separately.
- The practical impossibility of using a specialized simulation language because of the above difficulties (specialized languages are not adapted to handling event sets with 5000 events or more) and also because of the complexity of the structure; a network of 100 000 to 1 000 000 links requires storage and operations at the level of binary elements in order to economize memory capacity, which rarely permits a simulation language.
- The difficulty in measuring confidence intervals for the results because of the correlations between moments in time which are also far apart.

Management of the event set

As far as management is concerned (it is necessary to classify the elements in order to extract that with the minimum time), classical simulation studies compare three groups of techniques (cf. references).

- The linear event set which is a totally ordered table; if t_i denotes the time of occurrence of the event with position i, then $t_1 < t_2 < \ldots < t_i < t_{i+1} < \ldots$.
- The "pyramid"; the order here is partial as given by $t_i < t_{2i}$, $t_i < t_{2i+1}$, and hence a structure with pyramid order has $\log_2 N$ levels (where N is the number of stored objects). This results in a gain in time for the classification as soon as $N > 100$.
- The multiple-pointer event set which is a linear table with intermediate entry points which are themselves classified.

The performances of these algorithms have been compared (see for example [McCormack 81]). One conclusion from these comparisons is that the last two structures can be used in practice (the handling time increases as the logarithm of the number of classified objects), the pyramid presenting the advantage of not requiring dynamic optimization of the number of pointers.

Simulation by roulette

Simulation in real time requires as we have just seen the management of an event file, i.e. the loss of time (handling, and generation of event times) and loss of memory space.

When systems are handled which are entirely governed by exponential distributions, their properties make it possible to use another approach, simulation by roulette. An elementary example will serve to illustrate the principle.

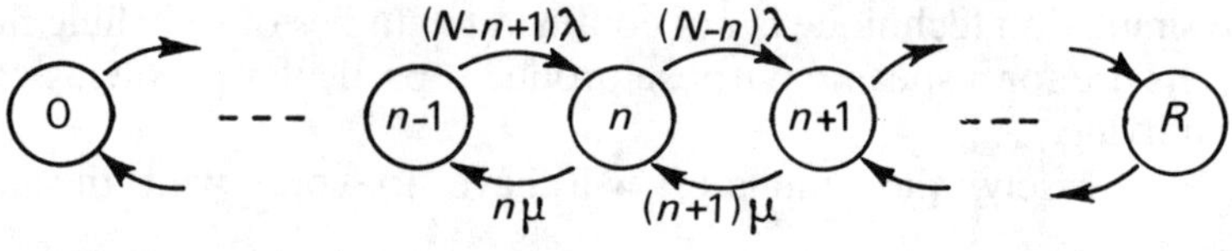

Fig. 4.12 Roulette simulation of Engset traffic.

Let it be necessary to simulate the behavior of a group of N sources in front of R servers (the Engset problem) under the assumption of exponential occupation distributions; the interval between requests coming from idle sources is also exponential. Figure 4.12 summarizes the process equations (cf. Chapter 2).

Let us suppose that the system is in the state n. It will leave it with certainty to pass into the state $n-1$ (one departure) or the state $n+1$ (one arrival).

The probability that the next change will take it from n to $n-1$ is

$$P(n \to n-1) = \frac{n\mu}{n\mu + (N-n)\lambda}$$

$$= \frac{n}{n + (N-n)\alpha}$$

where $\alpha = \lambda/\mu$. Similarly,

$$P(n \to n+1) = \frac{(N-n)\alpha}{n + (N-n)\alpha}$$

The program for simulation by roulette can be written in the following manner.

(1) Choose a random number r uniformly distributed between 0 and 1.

(2) Is r less than $n/\{n+(N-n)\alpha\}$? If it is, the event is a departure. Otherwise, it is an arrival. In both cases, the counts are updated (in particular, n).

(3) If the required number of events has not been achieved, return to (1).

The rejection rate is observed by counting the proportion of "arrivals" in the state $n = R$.

We emphasize that this method which reproduces the progress of the state is only permissible because of the exponential assumptions; this is again the "memoryless" property of the exponential.

In comparison with simulation in real time, the gain is derived from the economy of sampling based on the distributions and the economy of handling; no event file need be maintained with a resulting gain in time and memory.

The major disadvantage is the absence of any concept of time. The consequences of this are that it is impossible to handle systems with nonexponential time and that time (e.g. waiting time) cannot be measured.

This obstacle can be partly overcome by modifying the sampling process. K is chosen to be sufficiently large ($K > R + N$) (Fig. 4.13).

We then take a number between 1 and K. If sampling is made at ① in Fig. 4.13, it represents an arrival and, at ② in Fig. 4.13, it is a departure; otherwise it is an ineffective sampling which is abandoned.

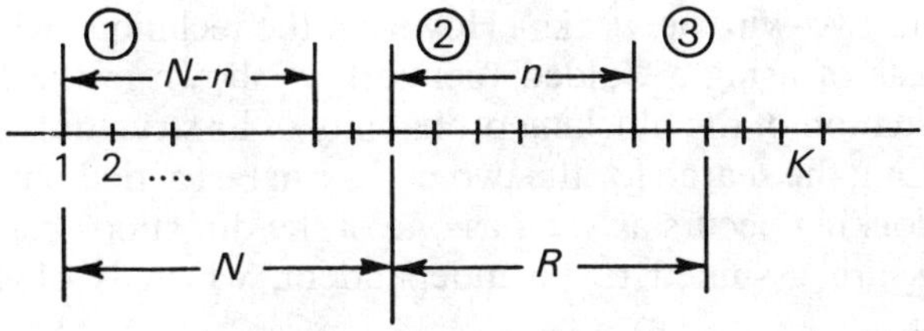

Fig. 4.13 Roulette simulation: time measurement.

In the Markov system which is simulated, the interval between events (of types ①, ② or ③) has a density

$$f(t) = K\,\mathrm{e}^{-Kt}$$

The number of samplings during the interval to be evaluated is counted. Let p_r be the proportion of observations where r numbers have been taken during the interval. The probability density for the duration of this time interval is then estimated from

$$p(t) = \sum_{r=1}^{\infty} p_r Kt^{r-1} \frac{\mathrm{e}^{-Kt}}{(r-1)!}$$

In particular, the mean is given by

$$E(T) = \frac{1}{K} \sum_{r=1}^{\infty} r p_r$$

The problems of accuracy

The simulation of connecting networks suffers from the same weaknesses as all other such experiments; it is a statistical test which provides a random answer whose accuracy has to be estimated. Various methods are available to the user for calculating a confidence interval: replication of the runs, spectral analysis, regenerative simulation and the block method (see references and in particular [Fishman 78]). It should be recognized that none of these methods is really satisfactory because of the acute problems of autocorrelation. The high cost (execution time) prevents any complete implementation of the methods described above.

4.7 Digital networks; folded networks

All the methods described until now are obviously general. The specificity of most digital networks derives from their four-wire structure. It is possible to construct a network by placing two identical levels in parallel (which immediately brings us to two-wire networks). However, the technique which is currently adopted consists of using a "folded" network as shown in the diagram in Fig. 3.16. The evaluation of the blocking probability of its structures depends on the control algorithm; the search for the two paths can be carried out independently. In this case, blocking occurs as soon as one of the directions has no path. If the two directions are assumed to be independent, which is obviously only an approximation,

$$P\,(\text{at least one way blocked}) = 2B - B^2 \simeq 2B \tag{4.29}$$

If the two ends are on the same matrix (Fig. 4.14), they will meet congestion as soon as less than two simultaneous paths are available (eqn (4.17)).

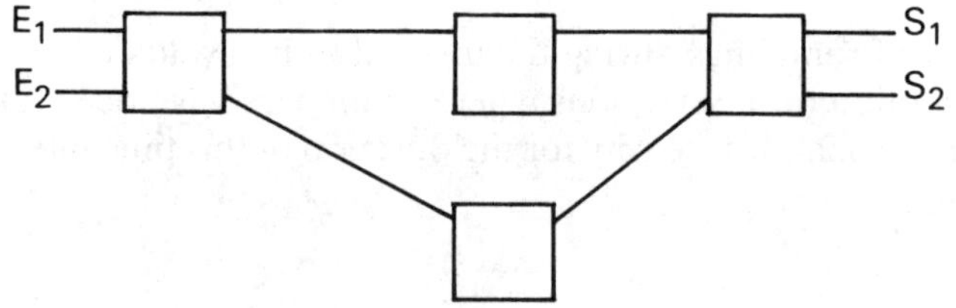

Fig. 4.14 Blocking in a folded network.

It is more advantageous, at the control level, simultaneously to search and assign the paths using a rule guaranteeing the availability of the second if the first is free; this is symmetric assignment already mentioned in Chapter 3.

The simplest rule discussed in [Gallassi 79] consists in using the same central matrix (most certainly in a three-stage network). For the requirements of a matrix towards itself, two independent paths will be used; if $r \geqslant n$, the blocking is zero (this can easily be seen since two internal links at least are idle on the matrix; by symmetry, the homologous links are idle between the central stage and the homologous matrix of the third stage).

It then remains to calculate the blocking between different matrices. The approach depends on the architecture. A combinatorial logic specific to each case may sometimes provide the "exact" value of the blocking. In general, it is difficult to take the symmetry into account. It can for example be "forgotten" by treating the two directions as independent and applying eqn (4.29) without always knowing the sense of the error which has been made. It is necessary to take care that the links are made busy and become idle two by two. The total traffic (in the sense of cumulative occupation of the links) is also doubled, each successful call intention occupying two paths.

Let us, for example, search for a path on the three-stage network in Fig. 4.15 with $r = n$ and used as a folded network. It is necessary to find a path from the two input matrices to the same central matrix.

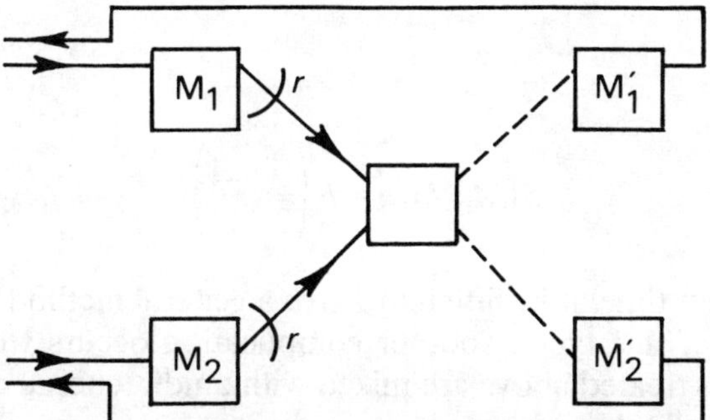

Fig. 4.15 Blocking in a folded network.

$$
\begin{aligned}
P\,(\text{blocking}) = {} & P(M_1 \text{ may reach 0 matrices}) \\
& + P(M_1 \text{ reaches 1 matrix})\, P(M_2 \text{ does not reach this matrix}) \\
& + \ldots \\
& + P(M_1 \text{ reaches } j \text{ matrices})\, P(M_2 \text{ does not reach these matrices}) \\
& + \ldots
\end{aligned}
$$

The calculation in Sections 4.2 and 4.3 is reproduced exactly and, for example,

$$B = \{a + (1 - a)\, c\}^{r-1}$$

under the *ad hoc* assumptions, a and c designating the traffic on the AB links.

The intermatrix links frequently consist of a "trunk group" of $f = 32$ or 64 links. The Lee method is then applied, replacing a term such as a by a^f (and correlatively by applying the Legall correction).

In other cases (TT structures in particular) the problem is simply solved (Fig. 4.16). The trunk groups between M_1 and M_2' accept connections which take the links two by two:

$$B(M_1, M_2') = E(f, A)$$

We ignore the path M_2 to M_1', which is always idle by symmetry. In contrast, for the connections of M to itself, the two connections always take a single trunk

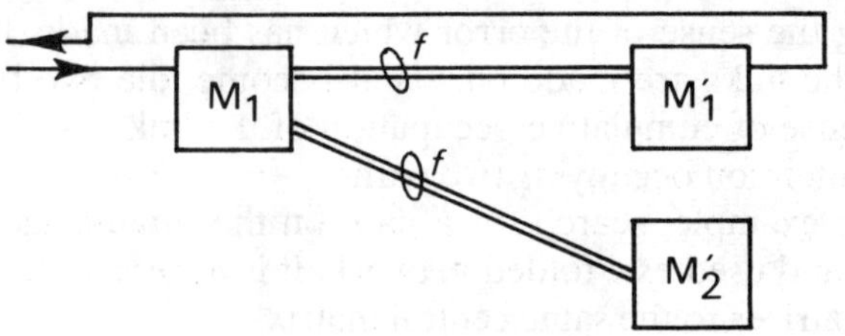

Fig. 4.16 Blocking in a *TT* folded structure.

group:

$$B(M_1, M_1') = E\left(\frac{f}{2}, \frac{A}{2}\right)$$

At the present time, it is difficult to give a general method of evaluation of folded networks of this type. Another complication occurs when the bidirectional connections treated above are mixed with unidirectional connections (e.g. connection of a sender to a circuit).

This is one of the tasks which will face evaluators of connecting networks, bearing in mind the development of this type of structure.

References

A large number of publications have been devoted to traffic calculation in switching networks. We list here the most important and useful references, most of which contain an additional bibliography.

The methods developed in this chapter were first presented in [Jacobeüs 50], [Lee 55] and [Legall 57]. These are "historic" papers. In [Elldin 56], it is proved, on the contrary, the practical impossibility of writing and solving state equations (Markov chain) for a network of typical size.

The papers below illustrate and extend the methods of the chapter given in [ESS1 79]. [Ess4 79], [ITT1250 81] and [Rothmaier 81].

[Bazlen 74] D. BALZEN, G. KAMPE, A. LOTZE. Design parameters and loss calculation in link systems. *IEEE Transactions on Communications,* Vol. COM 22, No. 12, December 1974, pp. 1908–1920.

[Bear 76] D. BEAR. *Principles of Telecommunication Switching,* Peter Peregrinus, London, 1976.

[Benes 65] V. BENES. *Mathematical Theory of Connecting Networks and Traffic,* Academic Press, New York, 1965.

[Benes 78] V. BENES. Reduction of network states under symmetries. *Bell System Technical Journal,* Vol. 57, No. 1, January 1978, pp. 111–149.

[Bininda 67] N. Bininda, G. Daisenberger. Recursive and iterative formulae for the calculation of loss in link systems. *Proceedings of the 5th International Teletraffic Congress,* New York, 1967, pp. 318–326.

[Elldin 56] A. Elldin. On equations of state for a two-stage link system. *Ericsson Technics,* No. 1, 1956, pp. 61–104.

[ESS1 79] R. A. Farel. Analysis and field results for No. 1 ESS network. *IEEE Transactions on Communications,* Vol. COM 27, No. 1, January 1979, pp. 10–15.

[ESS4 79] T. J. Cieslak, P. J. Hickson. Analysis and simulation of No. 4 ESS network. *IEEE Transactions on Communications,* Vol. COM 27, No. 1, January 1979, pp. 1–9.

[Feller 71] W. Feller. *An Introduction to Probability Theory and Its Applications* (2 volumes), John Wiley, New York, 1971.

[Gallassi 79] G. Gallassi. Blocking in 3-stage digital switching networks. *Proceedings of the 9th International Teletraffic Congress,* Torremolinos, 1979.

[Grantges 64] R. F. Grantges, N. R. Sinowitz. NEASIM: a general purpose computer simulation program for load-loss analysis of multistage central office switching networks. *Bell System Technical Journal,* Vol. 43, No. 3, 1964, p. 965.

[ITT1240 81] J. R. de los Mozos, A. Bucheister. ITT1240 Digital Exchange. Traffic Handling Capacity. *Electrical Communication,* Vol. 56, Nos 2–3, 1981, pp. 207–217.

[Jacobeüs 50] C. Jacobeüs. A study of congestion in link systems. *Ericsson Technics,* No. 48, 1950.

[Kodaira 76] K. Kodaira, K. Takagi. A general purpose blocking probability calculation program for multi stage link system. *Proceedings of the 8th International Teletraffic Congress,* Melbourne, 1976, Paper 331.

[Lee 55] C. Y. Lee. Analysis of switching networks. *Bell System Technical Journal,* Vol. 34, No. 6, 1955, pp. 1287–1315.

[Legall 57] P. Legall. Blocking calculation methods in conditional selection switching systems (in French). *Annales des Télécommunications,* Vol. 12, No. 11, 1957, pp. 374–386; Vol. 13, Nos 7–8, 9–10, 11–12, 1958.
Also *Proceedings of the 3rd International Teletraffic Congress,* The Hague, 1958.

[Lotze 77] A. Lotze, A. Roder, G. Thierer. PPL: a reliable method for the calculation of point to point loss in link systems. *Proceedings of the International Conference on Communication,* Chicago, June 1977.

[Pippenger 75] N. Pippenger. On crossbar switching networks. *IEEE Transactions on Communications,* Vol. COM 23, No. 6, June 1975, pp. 616–659.

[Rothmaier 81] K. Rothmaier, R. Scheller. Design of economic PCM arrays with a prescribed grade of service. *IEEE Transactions on Communications*, Vol. COM 29, No. 7, July 1981, pp. 925–935.

[Syski 86] R. Syski. *Introduction to Congestion Theory in Telephone Systems, Studies in Telecommunication,* Vol. 4, North-Holland, Amsterdam, 1986.

[Villar 72] J. E. Villar de Villacian, B. Fontana-Sanchis. Formulation of blocking functions for network graphs. *Electrical Communication*, Vol. 47, No. 4, 1972, pp. 239–244.

The major textbook concerning simulation techniques is [Fishman 78]. In [Weisschuh 73], [Villar 73] and [Dietrich 71] the special features of network simulation, especially roulette simulation, are discussed.

[Dietrich 71] G. Dietrich. Traffic model for the simulation of an entire exchange. *Electrical Communication,* Vol. 46, No. 2, 1971.

[Fishman 78] G. S. Fishman. *Principles of Discrete Event Simulation,* John Wiley, New York, 1978.

[McCormack 81] W. McCormack. R. Sargent. Analysis of future event set algorithms for discrete event simulation. *Communications of the Association for Computer Machinery,* Vol. 24, No. 12, December 1981, pp. 801–813.

[Villar 73] J. E. Villar de Villacian, J. R. de los Mozos Marques. An extension of the roulette model to simulate speech and signalling networks. *Proceedings of the 7th International Teletraffic Congress,* Stockholm, 1973, Paper 216.

[Weisschuh 73] H. Weisschuh. Theoretical aspects of finite source simulation. *Proceedings of the 7th International Teletraffic Congress,* Stockholm, 1973, Paper 215.

Chapter 5

PROBLEMS ASSOCIATED WITH TRAFFIC IMBALANCES

In order to carry out the calculations in the previous chapter, we are forced to make assumptions concerning the homogeneity of the sources and symmetry of the flows.

Each point of entry to the switching network (each source of calls) offers identical traffic and all this traffic is distributed homogeneously (i.e. there are no preferential streams between pairs of terminals).

The observance or the nonobservance of such a specification follows from the conflict between the following two phenomena.

- The differences intrinsic to the sources of traffic, circuits or subscribers, i.e. the difference in the volume of traffic and the existence of any privileged flows.
- The geometry of the switching network which must be as regular as possible and the path-searching algorithms which must handle all sources in the same manner, distribute loads uniformly and distribute the traffic streams (in order to attenuate the effects of inevitable inhomogeneities).

In the following, we study the effect of inhomogeneities of the sources, assuming complete symmetry of the network and the occupation algorithms (random choice of a route from all possible routes); the different strategies of choice will be discussed in the next chapter.

The concepts and calculations which are developed in this chapter assume their importance at the subscriber stages (subscriber line switching units and concentrators). It is this context which we shall consider (with the exception of Section 5.1 where we deal with group selection).

5.1 Existence of preferential flows

A switching network must distribute the various traffic flows. This is, as we recall, one of its essential functions. This distribution, if it is performed correctly, will be sufficient to cancel out any effects of disparities between the flows.

There is no doubt about the existence of these imbalances; a subscriber (the source of elementary traffic) only communicates with a few others (between 10 and 50 perhaps). A group of subscribers, subscriber line switching units or concentrators will certainly have nonisotropic, strongly individualized traffic. However, it can be assumed that the distribution imparted by the subscriber stage will be sufficient to eliminate the effects of this inhomogeneity provided that the connection from the subscriber stage to group selection is properly made.

Circuits are more difficult, especially under the assumption of digital transmission and switching. Let us consider, the most unfavorable case, the flow between "small trunk groups" consisting of a single PCM system, each subsequently connected to a single time-division matrix. Figure 5.1 illustrates the situation for small centers with a *TT* structure.

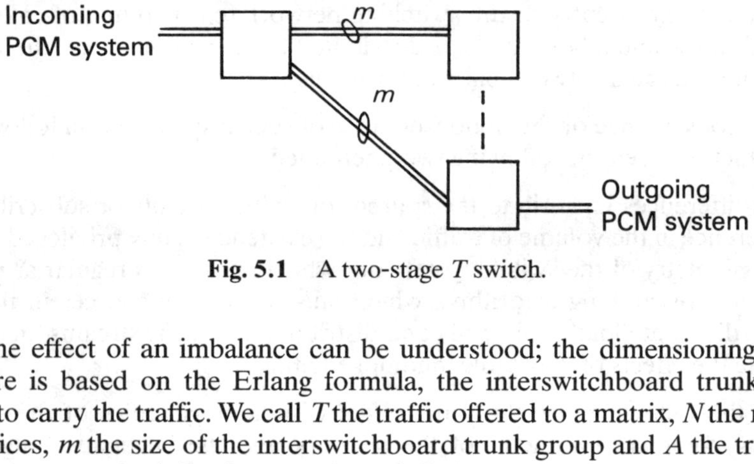

Fig. 5.1 A two-stage *T* switch.

The effect of an imbalance can be understood; the dimensioning of this structure is based on the Erlang formula, the interswitchboard trunk group having to carry the traffic. We call *T* the traffic offered to a matrix, *N* the number of matrices, *m* the size of the interswitchboard trunk group and *A* the traffic on this trunk group; the following must be satisfied:

$$E(m, A) \leqslant P \qquad (P = 1\% \text{ or } 0.5\%)$$

$$A = \frac{T}{N}$$

(the effect of "local" traffic, i.e. of the block on itself, is neglected).

The lack of mixing of the structure transforms the formula. Calling *u* the imbalance coefficient which is defined as the relative deviation with respect to the nominal value, it is necessary to verify that

$$E\left\{m, \frac{(1+u)T}{N}\right\} \leqslant P \tag{5.1}$$

The imbalance ($u \neq 0$) will be due to a large number of reasons: imperfect knowledge of the elementary flows between small trunk groups; intrinsic variability of the traffic; daily or seasonal variation of the traffic; effect of perturbations in the network (e.g. breakdowns); influence of overflow. The operator may correct these imbalances on average provided that it is possible to observe the internal links, which is lengthy and expensive (there are $N^2/2$ flows in a network with N matrices). However, it will not be possible to make slow variations disappear.

We emphasize the importance of the phenomenon which will affect "small trunk groups"; in a center of 2000 erlangs, the proportion of traffic carried by trunk groups of a single PCM system can be estimated as 40%.

The larger trunk groups which can be spread over several matrices will be less susceptible to the effects of imbalance.

There will be little that can be done to eliminate this phenomenon which is inherent in the technology used (e.g. it is impossible to spread a trunk group over several matrices but the use of a "deloader–decorrelator" of the type mentioned in Chapter 3 for ESS4 eliminates this phenomenon) and inherent to the structure "without distribution" of the two-stage networks (and more generally of the $2n$-stage time-division networks) (see [Hofstetter 76]).

The dimensioning must take this into consideration (see a possible estimation of u in Appendix to Chapter 5).

5.2 Inhomogeneous traffic sources

After having discussed the effect of nonhomogeneous flows, we shall examine the effect of the difference in volume of the source traffic. In practice, this is the subscriber case and, in one-sided networks, it involves the mixing of subscribers and circuits.

When the structure leads to mixing of circuits and subscribers, the existing imbalance will naturally be large, e.g. a circuit with 0.5–0.7 erlang and subscribers around 0.1.

In contrast, when only subscribers are involved, the reason for imbalances can be found in the differences in the telephone usage of the subscribers. To summarize, subscribers are classified in terms of their behavior.

- Residential subscribers who are private individuals using the telephone for their own usage produce a low traffic especially at peak time; their traffic varies from 0 to 0.05 erlang.
- Professional subscribers who consist of professionals, doctors, lawyers, artisans or small businesses with moderate usage of the telephone; their traffic varies up to 0.12 erlang.
- Business subscribers above 0.12 erlang for whom it is advisable to specify that they must be equipped so as to maintain their incoming occupation rate at

less than, say, 15%; this involves "line grouping", some group lines having traffic reaching or exceeding 0.5 erlang.

The problems arise from mixing different sources. This mixture may be voluntary and controlled; connecting a 0.7 erlang circuit to a matrix results in underequipping it with residential subscribers. It is thus important to know the rules, enabling the grade of service to be maintained for all users subjected to such a mixing. However, the variability within a single population may create considerable disorders (incorrect classification of a user to one of the categories, differences in usage etc.). Here, it will be important to be able to evaluate the effect of this "error".

In conclusion, there are two questions to be asked in this study.

- How is the traffic of a subscriber population distributed and what distribution is obtained by mixing users taken at random or carefully classified?
- What is the effect of these distributions on the traffic flow and on the grade of service experienced by each source?

The subscriber stages consist of matrices interconnected in concentration networks. The reply to the second question will be given in two stages: the effect of the imbalances in a matrix and the effect of the imbalances between matrices.

5.3 Imbalances in the subscriber traffic

We construct the following statistical model.

Each category of source (ordinary subscribers, business subscribers, circuit etc.) is considered as an infinite population characterized by a probability distribution

$$F_i(t) = P(\text{a source of category } i \text{ has traffic} \leqslant t)$$

and, when the sources are "mixed" by equipping a matrix of size N with subscribers (e.g. residential subscribers), it should be understood that N independent samplings are made from the population concerned.

This model shows how to reduce the question to a simple statistical problem. Again, it is necessary to have the functions $F_i(t)$ available.

In 1958, Hayward [Hayward 58] suggested an exponential distribution for F. He based this assumption on measurements carried out on a few concentrators. Figure 5.2 reproduces an example of the adjustments which he made.

It seems difficult to generalize these results indiscriminately. Hayward gives only a few details about the measurement; we only know that he observed seven "concentrators" for six consecutive weeks, each concentrator serving approximately 60 subscribers with average traffic 0.1 erlang.

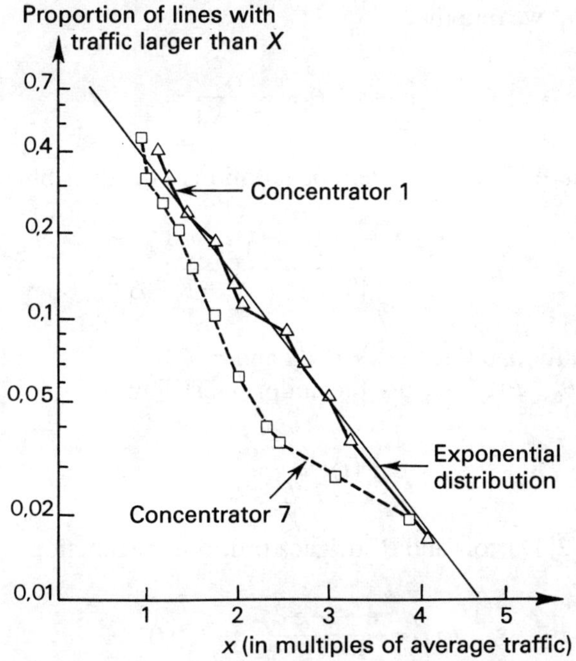

Fig. 5.2 Distribution of the subscriber traffic (according to [Hayward 58]).

Note that Fig. 5.2 reproduces the complementary experimental distribution $1 - F(t)$ for concentrator 1 (which is typical of these results) and for concentrator 7 (which is the only concentrator that if far from the proposed distribution).

Nevertheless, the measurements described are the only known example of such an investigation. We shall use it as a basis. Following Dartois and Rodriguez [Dartois 68], we shall assume an exponential distribution of the offered traffic, namely

$$F(t) = \frac{e^{-\theta_0/a} - e^{-t/a}}{e^{-\theta_0/a} - e^{-\theta_1/a}} \qquad \text{for } \theta_0 < t < \theta_1$$

Then,

$$\begin{aligned} F(t) &= 0 \qquad \text{if } t < \theta_0,\ a > 0 \\ F(t) &= 1 \qquad \text{if } t > \theta_1,\ a > 0 \end{aligned} \tag{5.2}$$

The average traffic α per subscriber will be given by

$$\alpha = a \frac{\theta_0\, e^{-\theta_0/a} - \theta_1\, e^{-\theta_1/a}}{e^{-\theta_0/a} - e^{-\theta_1/a}} \tag{5.3}$$

Setting $\Delta = \theta_0 - \theta_1$, we find that

$$\alpha = a + \theta_0 - \frac{\Delta}{e^{\Delta/a} - 1}$$

Similarly, the square of the coefficient of variation of the distribution is given by

$$c^2 = \left(\frac{\sigma}{\alpha}\right)^2 = \frac{a^2(e^{\Delta/a} - 1)^2 - \Delta^2 e^{\Delta/a}}{\{(a + \theta_1)(e^{\Delta/a} - 1) - \Delta\}^2}$$

For such a distribution, $0 \leqslant \alpha \leqslant (\theta_0 + \theta_1)/2$ and $c^2 \leqslant 1$.

When $\alpha = (\theta_0 + \theta_1)/2$, we use the uniform distribution

$$F(t) = \frac{t - \theta_0}{\theta_1 - \theta_0} \tag{5.4}$$

For $\alpha > (\theta_0 + \theta_1)/2$, Dartois and Rodriguez propose the function

$$F(t) = \frac{e^{\theta_0/a} - e^{t/a}}{e^{\theta_0/a} - e^{\theta_1/a}} \qquad a > 0 \tag{5.5}$$

which leads to

$$\alpha = \theta_0 - a + \frac{\Delta\, e^{\Delta/a}}{e^{\Delta/a} - 1} \tag{5.6}$$

Here also, $c^2 \leqslant 1$.

Since the distribution of the individual traffic of the sources obeys a distribution of the type in eqn (5.2), (5.4) or (5.5), it is easy to calculate the traffic offered to a matrix and therefore the imbalances appearing spontaneously for random assignment. The distribution of the traffic over a matrix receiving N subscribers will follow F_N:

$$F_N(t) = P(t_1 + t_2 + \ldots + t_N \leqslant t)$$

This is the convolution of N distributions identical with eqn (5.2), (5.4) or (5.5). Since the handling of the F_N distributions turns out to be rather laborious, Dartois and Rodriguez suggest a number of tables to enable a few simple problems to be solved. They also suggest approximate formulas for various examples. If $\theta_0 = 0$ and $\theta_1 = 1$, for example,

$$F_N(t) = 1 - \frac{P(N-1,\, t/a)}{E(N-1,\, t/a)} \tag{5.7}$$

where $E(N, A)$ designates the Erlang B formula and $P(N, A)$ the Poisson function. If N is sufficiently large, for example,

$$F_N(t) \approx \frac{1}{(2\pi)^{1/2}} \int_{-\infty}^{x} e^{-u^2/2}\, du \tag{5.8}$$

where

$$x = \frac{t - N\alpha}{\alpha N^{1/2}}$$

(this is the normal distribution).

The solution of the following problem provides an interesting application of the tables or the approximate formulas.

Given N sources taken at random from a homogeneous population with a mean α, what is the probability P that the overall traffic exceeds the value $N\alpha(1+u)$, i.e. has an imbalance coefficient u?

The curves in Fig. 5.3 give the imbalance obtained with probability P for $N=4$, 16 and 25, under the assumption of eqn (5.2) with $\theta_0 = 0$ and $\theta_1 = 1$.

In the region $\alpha \leqslant 0.16$, these curves show that the relationship only depends on N; Fig. 5.4 illustrates this dependence for $N=4$, 8, 16 and 25 (Figs 5.3 and 5.4 are taken from [Dartois 68]).

Figures 5.3 and 5.4 are sufficient in practice to solve the problems of characterizing "natural" imbalances in a homogeneous population of ordinary subscribers.

For a set of circuits where α is usually higher (0.6 erlang), the uniform distribution will be a better model. We use eqn (5.4) and

$$F_N(t) = \frac{1}{N!} \sum_{k=0}^{N} (-1)^k \binom{N}{k} \left(\frac{t - n_k}{\theta_1 - \theta_0} \right)^k$$

for $n_k < t \leqslant n_{k+1}$ $(k \leqslant N-1)$ and $n_k = N\theta_0 + k(\theta_1 - \theta_0)$.

As soon as $N \geqslant 4$, and for reasonable imbalances $(u \leqslant 1)$, the normal distribution is an excellent approximation to the above expressions. Then,

$$1 - F_N\{N\alpha(1+u)\} = P\{\text{total traffic} > N\alpha(1+u)\}$$

$$= 1 - \mathcal{N}\left\{ \sqrt{12N} \left(0.5 + \frac{\theta_0}{\Delta} \right) u \right\} \tag{5.9}$$

where $\mathcal{N}(x)$ designates the distribution of the reduced normal variable.

The form of eqn (5.9) makes it possible to plot universal curves giving P as a function of N and the parameter $u(\theta_0/\Delta + \frac{1}{2})$ (Fig. 5.5).

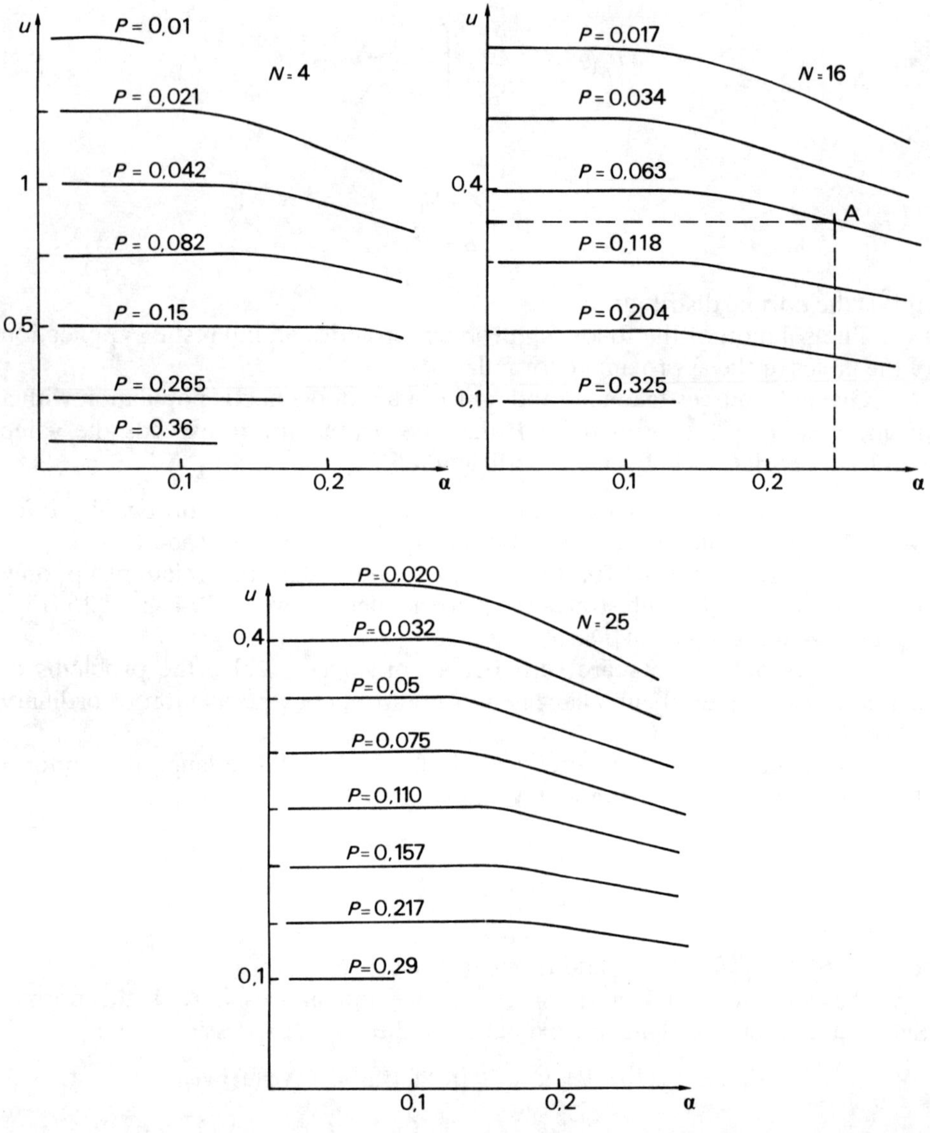

Fig. 5.3 Probability of exceeding the limit. The curves give the imbalance coefficient u as a function of N and for probability values P. For example (point A on the figure), a matrix receives 16 sources. The average traffic is 0.25 erlang. Therefore $P\{(\text{total traffic}) > (16 \times 0.25)(1 + 0.35) = 5.4 \text{ erlangs}\} = 6.3\%$.

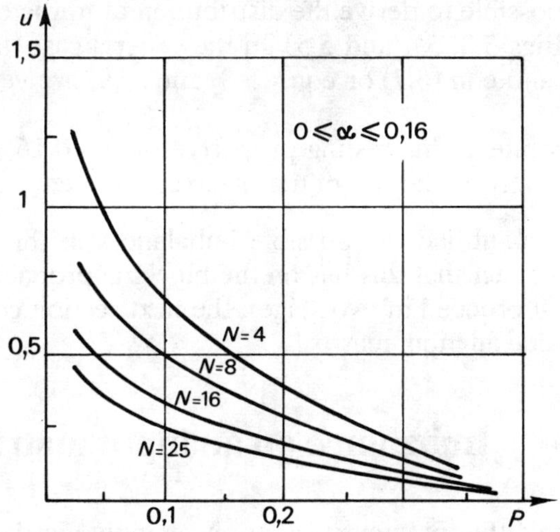

Fig. 5.4 Imbalance coefficient obtained with the probability P.

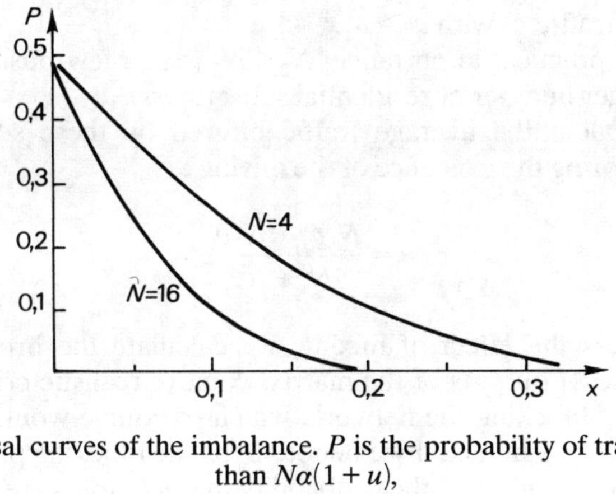

Fig. 5.5 Universal curves of the imbalance. P is the probability of traffic being greater than $N\alpha(1+u)$,

$$x = u\left(\frac{1}{2} + \frac{\theta_0}{\Delta}\right).$$

We shall briefly summarize the conclusions of this section.

• Equation (5.2), (5.4) or (5.5) provides a model for the distribution of a homogeneous population of sources. This model which is well suited to subscriber populations is only for coefficients of variation less than 1, which must cover the large majority of cases.

- From this, it is possible to derive the distribution of traffic originating from a matrix of size N (Figs 5.3, 5.4 and 5.5). In the general case, the distribution is difficult to handle and eqn (5.7) or eqns (5.8) and (5.9) are very good approximations.
- The curves illustrate an interesting property; for $a < 0.16$ erlang, the traffic imbalances depend only on the size of the group considered.

After having quantified the possible imbalances in this section, next we shall evaluate the impact that this has on the blocking probabilities. As already mentioned, we shall proceed in two stages; the next section concerns the effect of an imbalance inside an input matrix.

5.4 Imbalance on an input matrix

Traffic sources with different intensities supply an input matrix. It is not possible to give a quantitative result of universal application; we shall operate on particular schematic cases.

Let a matrix of size $N \times M$ receive two categories of sources (the generalization to k categories is obvious): N_1 sources offering traffic a_1 (low); N_2 sources offering traffic a_2 with $a_2 > a_1$.

In cases of practical importance, $N_2 < N_1$ (e.g. a few business subscribers mixed with a larger number of residential subscribers).

We shall call a the average traffic offered by these sources (which is evaluated by ignoring the existence of the mixing):

$$a = \frac{N_1 a_1 + N_2 a_2}{N_1 + N_2} \tag{5.10}$$

In order to assess the effect of mixing, we calculate the probability of total occupation of the M outputs of the matrix. A more realistic criterion would be the probability of blocking the network, which of course would depend on the nature of the network. In fact, if we denote B_i as the blocking probability for the source of category i, and $P_i(n)$ the probability that n outputs are busy, we have

$$B_i = P_i(M) + \sum_{n=0}^{M-1} P_i(n) P(\text{blocking} \mid M - n \text{ idle outputs})$$

$P_i(n)$ and $P_i(M)$ are related, and B_i and $P_i(M)$ exhibit a strong positive correlation.

It is possible to define several probabilities of matrix occupation: B_1 (or B_2) is the value observed by sources of type 1 (or 2), B is the weighted value, i.e. the value which is measured when the categories are mixed, and B^* is the value which would be observed if the sources were identical with offered traffic a.

B_1, B_2 etc, designate the call congestions perceived only by the users. We should point out the differences between these quantities. Let us assume that there is accidental mixing of sources; it is easy to verify and to balance the overall traffic of the matrix. The designer calculates the subscriber stage by assuming identical sources; he bases his dimensioning on B^*. An ordinary series of measurement will supply $\bar{B}$ (the percentage of rejected calls is observed). The users will experience B_1 or B_2 according to their category.

$\bar{B}$ is provided by the relationship

$$\bar{B} = \frac{N_1 a_1 B_1 + N_2 a_2 B_2}{N_1 a_1 + N_2 a_2}$$

B_1 and B_2 result from a calculation of the traffic, once the nature of flows 1 and 2 (Engset or Erlang traffic) has been defined. The calculation is facilitated by the existence of a "product form"; $P(i, j)$ denotes the probability of observing i calls of type 1 and j calls of type 2 under steady-state conditions; it is left to the reader to write the state equations using the techniques in Chapter 2. It can be verified from these equations that a solution exists of the form

$$P(i, j) = \text{constant} \times R(i)S(j)$$

For an Engset flow,

$$R(i) = \binom{N_1}{i} \alpha_1^{\,i}$$

$$S(j) = \binom{N_2}{j} \alpha_2^{\,j}$$

$$\alpha_k = \frac{a_k}{1 - a_k(1 - B_k)} \qquad k = 1, 2$$

The required call congestion can be derived from $P(i, j)$. Thus,

$$B_1 = \frac{\sum_i \lambda_1 (N_1 - i) P(i, M - i)}{\sum_{i,j} \lambda_1 (N_1 - i) P(i, j)}$$

The summation runs over all permissible states $0 \leqslant i \leqslant \min(N_1, M)$ and $0 \leqslant j \leqslant \min(N_2, M)$.

However, it is necessary to evaluate the constant; this can only be done numerically using the normalization equation. Figure 5.6 (taken from [Dartois 71]) shows the example of two business subscribers mixed with 14 ordinary subscribers on a 16×8 matrix as a function of the ratio $u = (a_2 - a_1)/a_1$.

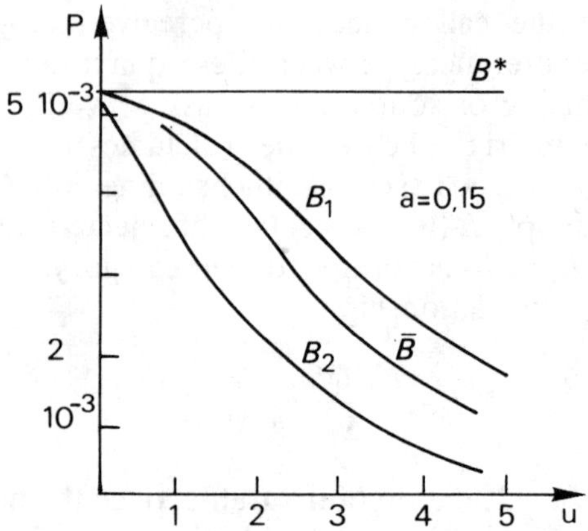

Fig. 5.6 Influence of imbalance in one matrix.

The favorable effect of mixing can be seen since B_2 and B_1 are appreciably lower than B^*. This phenomenon, which is astonishing at first glance, is due to the small number of business sources; these always see an idle output when they require it (two sources for eight outputs which are occupied little by sources with low traffic). At worst, sources with low traffic share six outputs, which is adequate for them. This beneficial effect will be observed as long as $N_2 a_2$ remains appreciably smaller than M.

The same phenomenon would be observed where subscribers and circuits are mixed. In contrast, when circuits are mixed with different traffic, carrying Erlang traffic, the reader can verify that $B_1 = B_2 = B^* \leqslant \bar{B}$ (the equality holds if and only if $N_1, N_2 > M$); in this latter case the B_i values are interpreted as congestion due to the network with the exclusion of any contribution made by the trunk group (when $N_i < M$). This is also due to the Poisson nature of the traffic (the verification of this is left as an exercise for the reader).

The following conclusions will be valid provided that the "business" sources are fewer than the matrix outputs.

- B_2 (business source) $< \bar{B}$ (weighted) $< B_1$ (low traffic source) $< B^*$.
- The greater the imbalance, the more significant is the effect.

The mixing of a few high-traffic sources (subscribers and circuits) with ordinary subscribers always has a beneficial effect. However, it is always difficult to quantify the gain exactly.

In [Dartois 71] the concept of gain in traffic capacity is defined and this gain is evaluated for various configurations. In the same way, he studied a complete network (a concentrator with two or three stages) and showed that the specified results remain valid.

The following points should be noted.

(1) In such a situation, it is worth making a clear choice of the criterion concerning the grade of service. The choice is made in the following way.

- The blocking B^* is calculated using classical methods. No use is made of the gain due to the imbalance.
- The blocking of the most underprivileged subscriber (with lowest traffic) must be guaranteed. The grade of service is good for everyone but will be expensive (particularly so because the subscribers who determine the dimensioning communicate to a small extent and thus make a small contribution to profits).
- The weighted blocking $\bar{B}$ is imposed; this solution may be a good compromise between grade of service and cost.

(2) The beneficial effect is obtained provided that the high-traffic sources do not saturate the outputs (the difference between the number of outputs and the traffic which they carry must remain higher than the traffic of low-weight sources; this condition will be met in practice).

5.5 Imbalances between matrices

It is understandable that a matrix may have total traffic of more or less than the nominal value.

Let us consider a switching network whose input matrices of size $N \times M$ receive subscribers with average traffic equal to a. Because of the imbalance, the traffic measured at any matrix will deviate from the average value Na; we call u the imbalance coefficient (the effective traffic is $Na(1+u)$). As previously, the effect of u can be evaluated qualitatively from the probability of total blocking $P_u(M)$ of the imbalance matrix u (Table 5.1).

Let us recall that in Section 5.3 we showed how to calculate the probability of such imbalances.

$P(M)$ does not give an exact picture of the effect of the imbalance. We therefore look at the example of the three-stage network in Fig. 5.7. For simplicity, we assume that inside one matrix the subscribers offer identical traffic.

The channel graph will be identical for all subscribers except for the traffic parameter of the AB links. The probability of internal blocking is given by eqn (4.5); the corresponding assumptions and approximations will be accepted as valid.

We denote B_0 (B_+ or B_-) as the blocking probability for a subscriber on a "normal" matrix (overloaded or underloaded). $\bar{B}$ and B^* have the same

Table 5.1 Values of the total blocking probability $P_u(M)$

16 × 8 matrix

a \ u	−0.5	0	+0.5	+1.0
0.05	0	3.3×10^{-7}	1.0×10^{-5}	6.0×10^{-5}
0.1	3.3×10^{-7}	6.0×10^{-5}	9.0×10^{-4}	5.0×10^{-3}
0.15	1.0×10^{-5}	9.0×10^{-4}	1.1×10^{-2}	5.0×10^{-2}

32 × 16 matrix

a \ u	−0.5	0	+0.5	+1.0
0.05	0	0	0	1.0×10^{-8}
0.1	0	1.0×10^{-8}	2.3×10^{-6}	1.0×10^{-4}
0.15	0	2.3×10^{-6}	4.0×10^{-4}	8.6×10^{-3}

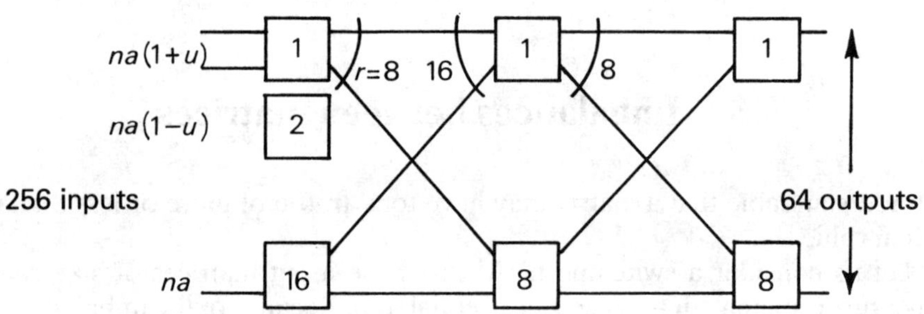

Fig. 5.7 Eight balanced matrices (four underloaded and four overloaded) are assumed.

meanings as previously ($B_0 = B^*$). $E(n, A)$ designates the probability of Erlang loss (n servers and offered traffic A). B_0, B_+ and B_- are given by

$$B_0 = \frac{E(r, C)}{E(n, C/a)}$$

$$B_+ = \frac{E(r, C)}{E\{n, C/a(1+u)\}}$$

and

$$B_- = \frac{E(r, C)}{E\{n, C/a(1-u)\}}$$

B_0, B_+, B_- and $\bar{B}$ values for $u = 0$, $+0.5$ and $+1.0$ are given in Table 5.2.

Table 5.2 B_0, B_+, B_- and $\bar{B}$ for various u values ($a=0, 1$)

u	B_0	B_+	B_-	$\bar{B}$
0	3.2×10^{-3}	—	—	—
+0.5	3.2×10^{-3}	5.2×10^{-3}	2.2×10^{-3}	3.7×10^{-3}
+0.1	3.2×10^{-3}	1.0×10^{-2}	3.2×10^{-3}	6.6×10^{-3}

It can be seen that an imbalance in the total traffic of the input matrices will always reduce the performances of the network. Depending on the configuration of the system, the imbalances will be manifested at several stages. We assume that the line switching unit is well balanced as a whole (i.e. that it carries its nominal traffic as a whole, which routine measurements readily guarantee). With the structure in Fig. 5.7, an input matrix may be overloaded, producing an imbalance in the *AB* links; the total distribution existing between the first and second stages guarantees a good balance on the *BC* links.

In contrast, the structure in Fig. 5.8 presents two possible levels of imbalance; the overload of the *K* matrices of a block has an effect not only on the *AB* links but also on the *BC* links since the different blocks are not distributed among themselves. The effect of this imbalance will be less serious, the large size of the subscriber group ensuring that it has a lower value.

As can be seen from the examples which can be multiplied to infinity the effect of the disparities between sources can only be examined in detail for a

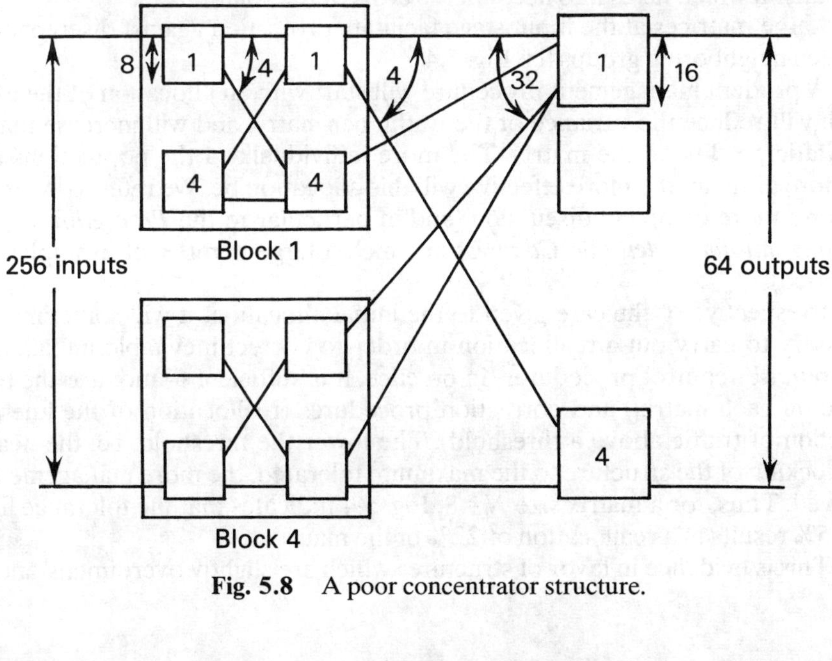

Fig. 5.8 A poor concentrator structure.

given structure. The choice of a structure must be influenced by considerations which govern the good flow of the traffic and ease of operation.

5.6 Conclusions; consequences for management procedures

As a first conclusion, we can say that, although there is no doubt about the effect of imbalances, their quantitative evaluation remains difficult and can only be carried out for each individual case.

The following qualitative statements can be made.

(a) The traffic imbalances between subscribers belonging to the same matrix reduce the blocking experienced by all subscribers.

(b) Imbalances between matrices increase this blocking; the closer the imbalance is to the input stages, the more marked is this effect.

In practice, attempts are made to incorporate these conclusions into the design and management of the subscriber stages.

If economic considerations are disregarded, it is better to mix different populations on the same equipment so as to profit most from rule (a) rather than using equipment adapted to each traffic range.

This conclusion, of course, will have to be moderated by an economic comparison which takes into account the costs of the solutions.

Large matrices at the input stage facilitate protection against discrepancies between neighboring groups (cf. Fig. 5.4).

A prudent management procedure will start with an allocation of the users which will reduce the variance of the traffic per matrix and will increase that of the traffic per line in the matrix. The more individualized the populations and the more of them, the more effective will this allocation be. We refer to [Morales 80] for a more complete discussion (and in particular to the *Proceedings of the 9th International Teletraffic Congress* in which a large number of examples are discussed).

Irrespective of the care given to the initial allocation, it will sometimes be necessary to carry out a reallocation in order to correct inevitable imbalances. This requires control procedures (in practice, it is sufficient to monitor the total traffic on each matrix) and correction procedures (reallocation of the lines on detection of traffic above a threshold). The lower the threshold, i.e. the nearer the blocking of the structure to the maximum tolerated, the more management is involved. Thus, for a matrix size $N = 8$, Fig. 5.4 indicates that the tolerance limit $u = 25\%$ results in a reallocation of 22% of the matrices.

This is evidence in favor of structures which are slightly overdimensioned.

Appendix to Chapter 5

Bell System engineers have been studying a phenomenon for a long time (see for example [Wilkinson 56], [Wilkinson 71] and [Hill 77]), which they have named "day-to-day variation" in the traffic.

Let there be a trunk group between exchanges (more generally, a traffic stream of the urban telephone network, the interurban telephone network etc.). The intensity of the traffic offered to the trunk group is measured and A is defined as the intensity of traffic at the busy hour. A can be measured on several consecutive days and then a dispersion is found in successive values which cannot be explained by the fluctuations inherent in any statistical experiment.

The measurement process consists of observing the trunk group for a given time. The occupation times can be totalled or sampling can be used; it is necessary to take into account the rejected calls. Of course, as in any measurement procedure, the estimate taken from the experiment is a random variable. We denote A as the mean of the traffic and $\hat{A}(T)$ as the estimate so that, if the offered traffic is Poissonian,

$$E\{\hat{A}(T)\} = A$$

$$\mathrm{var}\{\hat{A}(T)\} = \frac{2Ah}{T}$$

where h represents the average duration of service and T the observation time.

The measurements which have been carried out do not fit into this framework. In order to account for his results, Wilkinson was forced to assume that A is itself a random variable dependent on time; this is the day-to-day variation or rather the slow variation in the traffic. The explanation for this phenomenon could only be socioeconomic and of little interest here.

More precisely, the successively observed values are well represented by independent sampling of the variables obeying a gamma distribution:

$$P(A \leqslant x) = \frac{\beta^\alpha}{\Gamma(\alpha)} \int_0^x t^{\alpha-1} \, \mathrm{e}^{-\beta t} \, \mathrm{d}t \qquad \text{(A.5.1)}$$

with

$$\mathrm{mean}(A) = \mathscr{A} = \frac{\alpha}{\beta}$$

$$\mathrm{var}(A) = \frac{\alpha}{\beta^2}$$

and

$$\Gamma(\alpha)=(\alpha-1)\Gamma(\alpha-1)$$

(if α is an integer, $\Gamma(\alpha)=(\alpha-1)!$). In order to account for the dispersions, it is necessary to take

$$\mathrm{var}(A)=0.13\mathscr{A}^{\phi} \tag{A.5.2}$$

where ϕ varies between 1.5 and 1.85 depending on the case considered.

This model could be considered as representative of the slow fluctuations in the volume of any traffic stream. It is assumed that, inside an exchange in particular, the traffic between blocks in the configuration in Fig. 5.1 obeys a distribution of the type in eqn (A.5.1), the mean $\mathscr{A}$ taking the nominal value T/N (eqn (5.1)).

A numerical example will illustrate this idea and give the orders of magnitude.

Let us assume a center with average size (2000 erlangs), the average traffic between each block being 20 erlangs. In accordance with eqn (A.5.2), the variance of the day-to-day traffic will be

$$\mathrm{var}(\mathscr{A})=0.13\times 20^{1.85}=5.8^2$$

Consultation of a gamma distribution graph (see Appendix at the end of the book) tells us that the instantaneous traffic will lie between 10 and 33 erlangs in 95% of the cases.

A dimensioning criterion has to be established in order to use this model. For example the "maximum" blocking could be limited to correspond to the case in which the traffic A_{max} has a 95% chance of not being exceeded. In our example, $A_{max}=30$ erlangs.

The CCITT has adopted a somewhat different and less draconian standard for the dimensioning of trunk groups; the average observed blocking probability over a long period is defined as

$$\bar{B}=\int B(A)\,\mathrm{d}P(A)=B(A_{eq})$$

where $B(A)$ is the instantaneous blocking probability for the traffic A (e.g. the Erlang formula), $P(A)$ is the probability of observing a value less than A (eqn (A.5.1)) and A_{eq} is the constant traffic which would cause $\bar{B}$.

The resulting imbalance is

$$u=\frac{A_{eq}-A}{A}$$

It is obviously less than in the first case but the resulting dimensioning seems more realistic.

Recommendation E521 of the CCITT [CCITT 84] adopts this method and gives dimensioning tables for traffic averages up to 30 erlangs. The method of calculation is developed in [Hill 76].

References

Relatively few studies can be found on the influence of imbalances on the quality of service, and even fewer on case studies. The following references are probably an exhaustive bibliography on the subject. Most of the contributions have been made by ITT engineers (the Madrid laboratories especially).

The paper by Morales is a useful addition to Section 5.6. A statistical device is described in [Pedersen 76] which permits optimization of the load balance effort.

[CCITT 84] CCITT. *Red Book*, Vol. 2, Nos 2–3, 1984.

[Dartois 68] J. P. DARTOIS, A. RODRIGUEZ. Traffic unbalances in small groups of subscribers. *Electrical Communication*, Vol. 43, No. 2, 1968.

[Dartois 71] J. P. DARTOIS. Lost call cleared systems with unbalanced traffic sources. *Electrical Communication*, Vol. 46, No. 1, 1971.

[Hayward 58] W. S. HAYWARD. Traffic engineering and administration of line concentrators. *Proceedings of the 2nd International Teletraffic Congress*, The Hague, 1958.

[Hill 76] D. W. HILL, S. R. NEAL. Traffic capacity of a probability engineered trunk group. *Bell System Technical Journal*, Vol. 55, No. 7, September 1976, pp. 831–842.

[Hill 77] D. W. HILL. The effects of traffic load variation on measurement accuracy. *Bell System Technical Journal*, Vol. 56, No. 4, April 1977, pp. 561–573.

[Hofstetter 76] H. HOFSTETTER. The effects of non-uniform traffic distribution in switching networks. *Proceedings of the 8th International Teletraffic Congress*, Melbourne, 1976, Paper 342.

[Morales 80] G. MORALES, A. CRESPO, M. G. RAFART. Traffic imbalances in subscriber matrices. *Electrical Communications*, Vol. 55, No. 1, 1980, pp. 18–25.
Also *Proceedings of the 9th International Teletraffic Congress*, Torremolinos, 1979.

[Pedersen 76] O. A. PEDERSEN. Statistical design load balance traffic studies. *Proceedings of the 8th International Teletraffic Congress*, Melbourne, 1976, Paper 634.

[Wilkinson 56] R. I. WILKINSON. Theories for toll traffic engineering in the USA. *Bell System Technical Journal*, Vol. 35, No. 2, March 1956, pp. 421–454.

[Wilkinson 71] R. I. WILKINSON. Some comparisons of load and loss data with current teletraffic theory. *Bell System Technical Journal*, Vol. 50, No. 8, October 1971, pp. 2807–2834.

Chapter 6

SWITCHING NETWORKS; FINAL REMARKS

This chapter marks the end of the first part of this book which concerned the study of traffic flows in networks. The preceding chapters should not have given the idea that we are dealing with a perfected technique which is already somewhat out of date. Many problems remain open and the development of the technologies involved constantly raise new questions; some of these will be touched on here.

In Section 6.1, we shall deal with the evaluation of networks operated with delay. Practical application involves the subscriber line switching units and in particular preselection. A model with delay may reflect actual operation or may approximate reality. As we shall see, the resulting calculations, despite very favorable assumptions, do not yield very simple expressions.

Section 6.2 discusses comparative performances of rules for allocating paths as well as the rearrangement technique. The latter turns out to be very attractive (it reduces or eliminates blocking at a low cost of the control system). It remains to extend the application to more general classes of structure.

The theory of complexity is covered by Section 6.3. This is a brief survey aimed at explaining the objectives and the type of results to be expected and stimulating a more in-depth study.

Finally, Section 6.4 presents a thermodynamic approach of the statistical mechanics type which is very attractive and sheds new light on traffic phenomena in complex structures such as switching networks. This theory has certainly not realized all its potential.

The last two sections, even more so than the preceding sections, are simply an introduction to a more complete study; this explains the particular importance of the references. [Benes 65] is very stimulating reading and is especially recommended.

6.1 Link systems with delay

Most calculations of blocking probability in networks assume a loss system. In fact, this is at best a simplifying assumption; nothing stops a connecting network from operating "with delay" apart from the difficulties indicated below; in addition, the model with delay may be a reasonable approximation.

In fact, let us consider a local exchange in which the new calls are detected by the control system using a scanner (cf. Chapter 7). In this type of organization the scanner polls the subscribers in groups and detects changes in state. Unhooking initiates preselection; search for a dialing receiver unit which must be connected to the subscriber. The failure of the path search causes the call to be abandoned either by sending the busy tone to the caller who must hang up and make another attempt (this is a reattempt system for which a model with delay is an approximation) or by pure and simple abandonment, i.e. the call will be detected again in the next cycle and a new search made which may be terminated (in this case also, a model with delay is a perfectly reasonable approximation). Finally, the control system may keep the call in reserve in order to attempt to route it again as soon as a path is idle (a true system with delay). The last solution turns out to be costly in terms of processor time and not very effective.

The waiting time from the user's point of view will in any case be rather long; it is expressed as a fraction of the conversation time and is of the order of 1 min. In addition, when each path becomes idle in the network, the control system must try (most often in vain) to route one of the waiting calls; it is necessary to solve the problem of choice and to manage the collection of waiting requests. Consequently, there is an increase in the load on the control system, and in a congested period an even larger increase! The situation is completely different from that for the control units where waiting is a common fate. It is the complexity of the switching network which cancels out the advantages which could be derived from it. In summary, it is not certain that there is any point in operating the switching network with delay.

Nevertheless, in all these cases, the model with delay is a realistic approximation of the functioning of preselection and is perhaps more realistic than the traditional loss model.

It is necessary to define the disciplines; they will be quite different for each system but a model with service in random order seems to be completely plausible (the order of actual service will depend on a large number of design factors, operational factors etc.). On the assumption of a loss system operation, it would be sufficient to postulate a Poissonian arrival distribution. In this case, it would also be necessary to assume an exponential service time (dialing plus conversation). It then becomes necessary to answer the two following questions.

- What is the probability of experiencing waiting (failure in the path search)?
- What is the waiting distribution (or, at least, the average waiting time)?

Two possible approaches will be discussed here: a modification of the Jacobeüs method and the application of the effective availability method.

We shall present these two methods for point-to-point selection using a three-stage network as an example (Fig. 4.1).

6.1.1 The Jacobeüs method

This is a direct adaptation of eqn (4.1) which is rewritten with the same notation as in Section 4.2:

$$P(\text{waiting}) = \sum_{j=0}^{r} G(j)H(r-j) \tag{6.1}$$

$G(j)$ is the probability of observing j busy links and $r-j$ idle links, and $H(r-j)$ is the probability that the given $r-j$ links are busy.

G and H will be given by expressions corresponding to traffic operated with delay (cf. Section 2.9). For offered Erlang traffic (A erlangs per matrix and r outputs), they are

$$\begin{aligned} G(j) &= G(0)\frac{A^j}{j!} \qquad \text{if } j<r, \\ G(r) &= G(0)\frac{r}{r-A}\frac{A^r}{r!} = E_2(r, A). \end{aligned} \tag{6.2}$$

For Engset traffic (n subscribers with coefficient $\alpha = \lambda/\mu$ per matrix equipped with r outputs ($n > r$), they are

$$\begin{aligned} G(j) &= G(0)\binom{n}{j}\alpha^j \qquad \text{if } j<r \\ G(j) &= G(0)\binom{n}{j}\frac{(n-r)!}{(n-j)!}\frac{\alpha^j}{r^{j-r}} \qquad \text{if } r \leqslant j \leqslant n \end{aligned} \tag{6.3}$$

with

$$G(0) = \left\{ \sum_{k \leqslant r-1} \binom{n}{k}\alpha^k + \alpha^r \binom{n}{r} \sum_{k=0}^{n-r} \left(\frac{\alpha}{r}\right)^k \frac{(n-r)!}{(n-r-k)!} \right\}^{-1}$$

(The proof of eqn (6.3) is left as an exercise to the reader.) For Bernoulli traffic, of course, the same expressions as in Chapter 4 are found.

Expressions such as eqn (6.2) and, in particular, eqn (6.3) turn out to be rather cumbersome to handle. For example, when the basic assumption is made of Erlang traffic for two stages, we have for H, according to eqn (2.22),

$$H(r-j)=\sum_{k=r-j}^{r} G_2(k)\frac{\binom{j}{r-k}}{\binom{r}{k}}$$

$$=\frac{r-A+AE(j,A)}{r-A+AE(r,A)}\frac{E(r,A)}{E(j,A)} \tag{6.4}$$

E is given by the Erlang B formula (loss system). This is the Palm–Jacobeüs formula with delay. G_2 is the analog of G for the BC stage and is given by eqn (6.2). A denotes the traffic offered to the matrix of the first stage and C the traffic offered to the matrix of the third stage. The calculation is rather laborious; it gives

$$P(\text{waiting})=\frac{CE_2(r,C)(r-A)-AE_2(r,A)(r-C)}{(C-A)r}$$

except for $C=A$, where we have

$$P(\text{waiting})=[r+(r-A)^2+A\{1-E_2(r,A)\}]\frac{E_2(r,A)}{r}$$

As far as the waiting time distribution is concerned, eqn (4.1) can again be modified, this time to

$$P(\text{waiting time}>t)=\sum_x S(x)F(t,x) \tag{6.5}$$

where $S(x)$ is the probability that an incoming call finds the state x in the network. $F(t,x)$ is the probability of waiting longer than t for state x to be found. The summation runs over all the states where blocking is possible.

The function S is evaluated as above. A blocking state x will be referenced by (i,j), i designating the number of busy AB links and j the number of busy BC links:

$$S(i,j)=G_1(i)G_2(j)\frac{\binom{i}{r-j}}{\binom{r}{j}} \qquad \text{for } 0\leqslant i<r,\ i\leqslant j\leqslant r$$

In contrast, the evaluation of F turns out to be tricky; the approach in [Altamirano 78] involves separation of the events subjected to waiting to the position where the calls will be blocked and evaluation of H in each of these cases in a "very heuristic" way. The waiting states are split into three components depending on their cause: no AB links, no BC links or internal blocking (non-alignment). In the latter case, it is assumed that the blocked call is alone in facing this situation. This approximation is justified by the low proportion of communications between the two matrices (cf. Fig. 3.12 where it is shown that a situation of internal blocking involves only the pair of input and output matrices). The call waits until a link blocking becomes idle. There are $i+j-r$ such links. Let us call h the (exponential) duration of the calls. Then

$$F(t, i, j) = e^{-(i+j-r)t/h}$$

However, if the AB links are all busy, several inputs of the matrix will possibly be waiting; we are dealing with an $M/M/r/n$ system. Let us suppose that a link becomes idle. Let us call f the probability that a call may use it, i.e. the probability that the corresponding BC link is idle:

$$f = \frac{C}{r} \text{ (traffic per } BC \text{ link)}$$

The fate of a call will also depend on the service discipline; if the choice is made at random, it will finally be chosen and will have a path (possibly, after other inputs have been tried) with probability

$$p = \frac{1-f^j}{j},$$

where j is the number of other waiting calls. If it is not serviced, it must wait for the next departure. The calculation finally leads to rather cumbersome expressions. The interested reader should consult the above-mentioned reference. However, there is good agreement with simulations.

6.1.2 The effective availability method

This involves adapting the method presented in Chapter 4; the network is replaced by a single stage with restricted availability and the waiting probability is calculated for this.

As in Chapter 4, waiting is separated into two events and with k_{eff} denoting the effective availability we can write

$$P(\text{waiting}) = E_2(r, A) + \{(1 - E_2(r, A)\} H(r - k_{\text{eff}}) \quad (6.6)$$

The function H is given directly by eqn (6.4). The effective availability k_{eff} is calculated as in Chapter 4. Naturally, $E_2(m, A)$ is given by the Erlang C formula.

This approach also leads to the waiting distribution. As in determining eqn (6.6), the events subjected to waiting are separated to the position where blocking is produced, and the total waiting time is the sum of the waiting time in front of the AB links and the internal network, i.e. in front of the "grading" which replaces this.

We call $W_{\mathrm{IDF}}(t)$ the waiting distribution in this latter case; it is calculated assuming an idealized system, and whence the name "ideal grading formula"; this system has been studied in [Haugen 77] and also by others:

$$P(\text{waiting time} > t) = W(t) = W_E(t) + \{1 - W_E(0)\}\, W_{\mathrm{IDF}}(t)$$

$$+ \int_{0+}^{t} W_{\mathrm{IDF}}(t-u)\, \mathrm{d}W_{\mathrm{E}}(u) \qquad (6.7)$$

$W_E(t)$ is the waiting distribution in front of the AB links:

$$W_E(t) = 1 - E_2(m, A)\, \mathrm{e}^{-(m-A)t}$$

(the service time is taken as equal to unity and the calls are assumed to be serviced in the order of their arrival).

The average waiting time is derived from eqn (6.7):

$$E(T) = T_{\mathrm{IDF}} + \frac{E_2(m, A)}{m - A}$$

T_{IDF} is the average of the waiting time before the grading. T_{IDF} and W_{IDF} are given by approximations, the simplest consisting of replacing W_{IDF} by an Erlang distribution whose parameters are obtained by adjusting the moments:

$$W_{\mathrm{IDF}}(t) \approx \int_{\lambda t}^{\infty} \mathrm{e}^{-y} \frac{y^{p-1}}{(p-1)!}\, \mathrm{d}y$$

with

$$\frac{T_{\mathrm{IDF}}}{W} = \frac{p}{\lambda}$$

$$1 + \frac{1}{p} = 4\left\{2 - \frac{A}{n}(1 - M^{-2/3})\right\}^{-1}$$

where

$$M = \binom{n-1}{k-1}$$

The function $\sigma(x)$ is introduced to give the probability of blocking the grading with availability k in the state x; this is the probability that the k accessible outputs are blocked, if x are blocked as a whole:

$$\sigma(x)=\frac{\binom{x}{k}}{\binom{n}{k}}$$

$p(x)$, which is the probability of finding that j outputs are busy, is given by

$$p(x)=p_0A^x\frac{\prod_{i\leqslant x-1}\{1-\sigma(i)\}}{\sum_{i=1}^{x}\{1-\sigma(i)A\}}\qquad 0\leqslant x<n$$

(p_0 is obtained by normalization.) From this, the waiting probability before the grading is derived:

$$P(\text{waiting})=\sum_{j=k}^{n}p(j)\sigma(j)$$

We set

$$w(x)=\sum_{i=k}^{x}\frac{A\sigma(i)}{1-A\sigma(i)}$$

$$\Omega=\sum_{j=k}^{n}p(j)w(j)$$

Ω is the average length of the queue from which we derive the average time required using the Little formula:

$$T_{\text{IDF}}=\frac{\Omega}{A}$$

As can be seen, the application of this method is also far from easy; the "proofs" given only justify it partially. Nevertheless, here also, comparison with simulation gives some support to the approximations.

6.2 Path selection algorithms; rearrangement

In Chapter 3, we indicated the possibility of constructing different algorithms characterized by the choice of a path among all those available. In Chapter 4, it was necessary to assume a random choice. In this section, we shall examine the influence of other choices. We shall illustrate these considerations by examining the simple case of the symmetric Clos network (r central $b \times b$ matrices and nb inputs).

The first method of choosing a path (after having searched for all possible paths using the procedure in Section 3.3) uses a sequential rule; search for a path starting from the same initial point and stop at the first path found. We thus scan matrix 1 of the central stage, then matrix 2 etc. In order to minimize the effect of nonuniform distribution of wear (for space-division networks) or of concealed breakdowns, it is possible to change the point of departure periodically. Processor control can easily be adapted to such an algorithm.

Another rule, which is more costly to implement, involves searching for a path taking the most heavily loaded matrices. This rule is called the "call-packing (or saturation) strategy". It is obvious that it eases traffic flow because it preserves empty matrices for difficult cases in the best possible manner. The search for the matrix with the heaviest load may involve a nonnegligible calculation time. In this respect the previous strategy gives similar results at lower cost.

A variant of the call-packing strategy has been suggested which has been named "call repacking" [Ackroyd 79] and which is based on the observation that the saturation only partly manages to perform its function (saturation of the matrices); more paths become idle in full matrices than in the others. Furthermore, a weakness in the saturation strategy is that it requires the control system during the establishment of the call at a time when the processor time is costly. The repacking involves "stacking" the paths already established on each end of communication. To establish a path, it is chosen according to a sequential strategy (from 1 to r). Each time that a matrix of rank j becomes idle, a search is made for a call, using a matrix of higher rank which can replace it; it is then necessary to carry out a rearrangement.

Repacking is not an urgent task; it does not burden the control system as above (it can be delayed without congestion). It is even well adapted to a distributed control system.

We have spoken of rearrangement. This involves displacing a call which is already established. This procedure is used more systematically in the rearrangement strategies discussed later (repacking can be regarded as a preventive rearrangement).

Rearrangement may cause some problems in terms of equipment, in particular for networks with space-division technology. In time-division technology it is only a matter of modifying an address in the memory without interrupting the electrical connection. However, the result is an overload on the control units of the network (the "markers").

How is it possible to compare the respective merits of the above-mentioned strategies? This can be done by comparing their efficiency (i.e. the probabilities of blocking which they produce) and their costs (i.e. the control processor time which they need). The order of increasing efficiency for the above-mentioned strategies is as follows.

(a) Random search (on average, the least expensive in terms of calculation time).
(b) Sequential search (the simplest to implement).
(c) Saturation strategy (the most expensive).
(d) Call-repacking strategy.

This classification, which is widely accepted, results from comparisons by simulation (see for example [Bazlen 73], [Bonomi 80] and [Bonomi 81]). In fact, it is only known how to calculate the performances for the random search strategy.

As an example, in [Ackroyd 79] the call-repacking and call-packing strategies (the first divides the blocking probability by a factor which can be as high as 10) are compared. Figure 6.1 (taken from [Ackroyd 79]) illustrates the comparison for a three-stage network ($n=r=b=10$).

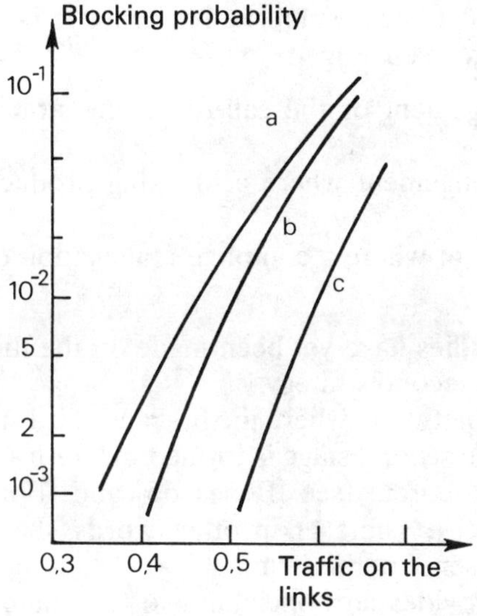

Fig. 6.1 Comparison of routing strategies: curve a, random search; curve b, sequential search; curve c, call-repacking strategy.

All these strategies have been described within the framework of the three-stage network. They can be immediately generalized to structures of the same type, a self-similar network in which the central matrix is replaced by a block formed by a three-stage network; the strategy is applied iteratively here by choosing the central block and then a path within it. In the case of structures which are usually used, the definition of the analog of these strategies is not obvious. Furthermore, the differences mentioned above become much less.

In general, the differences are also reduced when the traffic increases, all strategies becoming equivalent above 0.7 erlang per link.

Finally, it should be noted that no study has been carried out which seriously quantifies the relative costs (i.e. the time consumed).

To the author's knowledge, only the random strategy is currently being used.

The following points should be noted.

(1) In a two-stage connecting network, the preceding conclusions no longer apply. This is because the dimensioning rules are specific to this case owing to the absence of a distribution (cf. Section 5.1). On the contrary, these networks have the advantage of making the traffic uniform and the random strategy is the best in this case [Hofstetter 72].

(2) For nonblocking Clos networks, the application of rules (a), (b), (c) or (d) does not modify the limit; the condition $r > 2n-1$ must be satisfied irrespective of the strategy in order to guarantee strict nonblocking.

We shall now describe the rearrangement strategies which involve changing the routing of calls already established in order to be able to accept others. Several procedures can be envisaged.

- Preventive rearrangement or the call-repacking strategy already described above.
- Step-by-step rearrangement where a blocking produces a localized reconfiguration.
- Global rearrangement where a complete reallocation of paths is carried out periodically.

No important studies have yet been made on the third strategy. Here, we shall consider only the second strategy.

For a three-stage network, where the first and third stages are formed from b $n \times r$ matrices and the second stage is formed from r $b \times b$ matrices with $r \geqslant n$, the Slepian–Duguyd theorem (see [Benes 65] and [Bonomi 80]) states that rearrangment is always possible or in other words, that the network is nonblocking (in the broad sense of the term).

In addition, it provides an upper limit to the number of rearrangements which will displace $2(b-1)$ calls at worst ($b-1$ iterations in the algorithm below).

The theorem provides an explicit algorithm for the rearrangement. We shall present it by introducing a fruitful method of representation for this type of problem [Bonomi 81]. The Clos network is represented by Fig. 6.2.

A communication between matrix i of the first stage and matrix j of the third stage taking the central matrix as k will be represented by the symbol j appearing in the slot (k, i) (kth row and ith column).

The rules for operating the connecting network are expressed by the constraints of handling the table.

(a) A number j appears, at most, once on each row (the central matrix k only has one connection with a given matrix j).

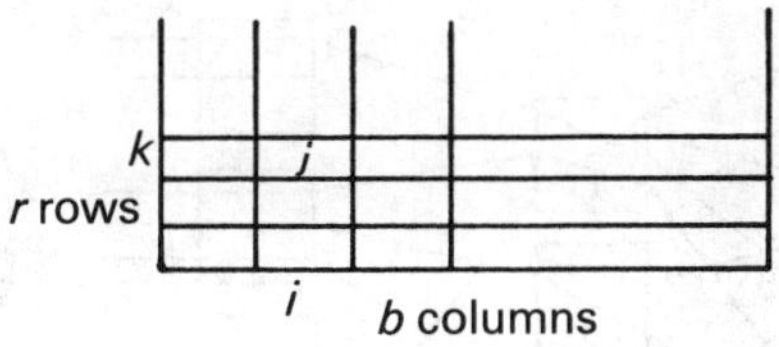

Fig. 6.2 Symbolic representation of the network.

(b) A symbol j appears, at most, n times in the table (there are, at most, n symbols per column).

Let us assume that the call (h, m) between the input matrix h and the output matrix m is blocked. On the matrix m, at most, $n-1 < r$ links are busy; there therefore exists a central accessible matrix. Let p be this matrix; the slot (row p and column h) of the table is empty.

Let L be the row of rank p of the table. If we place the call (h, m) in the slot h, constraint (a) will be violated; another slot, k, already contains the symbol m:

$$L = (d_1 \ldots, d_h = m, \ldots, d_k = m, \ldots, d_b)$$

There exists (at least) one row L' (of rank q) where the symbol m does not appear since the matrix m possesses at least one idle BC link:

$$L' = (w_1, \ldots, w_h = s_1, \ldots) \qquad w_1, w_2, \ldots \neq m$$

We set $y_1 = h$, $w_{y_1} = s_1$. Then, we have the following.

(1) If $j = y_1$ and $d_j = s_1$, we permute the slots d_{y_1} and w_{y_1} which unblocks the system (this means that the call from y_1 to m takes the central matrix p, and the call (h, m) the matrix q).

(2) If this is not the case, let y_2 be the slot such that $d_{y_2} = s_1$. We have unblocked (y_1, m) but blocked (y_1, s_1).

The procedure is reiterated replacing y_1 by y_2 and returning to (1).

Since the rows L and L' are not identical, we necessarily find a y_s value satisfying (1) or such that $w_y = \phi$. The initial row has a length b and we shall have made at most $b-1$ permutations.

As an example, consider the network in Fig. 6.3 which is represented by the table to the right of Fig. 6.3. Let it be necessary to route the call (II.1) there We place the call on row 1 (broken lines in the network). The successive rearrangements lead to the following table:

4	1	1	2		4	2	1	2		4	2	1	—
—	1	—	2		—	1	—	2		—	1	—	2
3	2	4	—		3	1	4	—		3	1	4	2

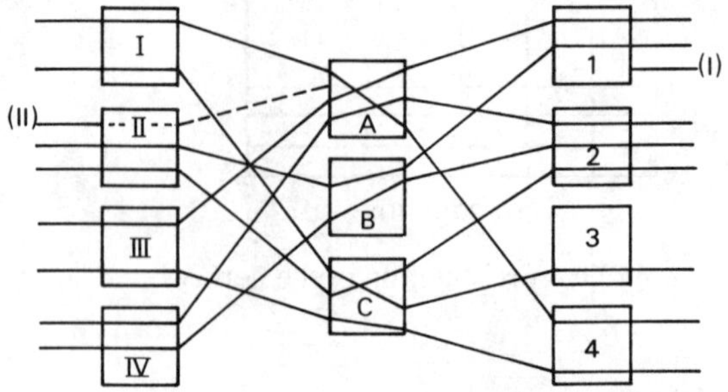

	I	II	III	IV
A	4	-	1	2
B	-	1	-	2
C	3	2	4	-

Fig. 6.3 Rearrangement in the network.

The last rearrangement is permissible, all the calls being processed. The reader can transpose the table to the diagram in Fig. 6.3 and check that it corresponds to the problem.

Note that the rearrangement strategy makes it possible to construct a three-stage network in which $n=r$ nonblocking; however, it is possible to associate the rearrangement with any of the allocation strategies (random, sequential or saturation). This choice will then modify the frequency and the scope of the rearrangements.

Thus the theoretical limit ($b-1$ displacements) has very little practical importance. Studies on actual cases show that the average number of rearrangements to be carried out does not exceed 0.1–1 for the usual range of traffic (see [Bonomi 83] and [Jajszczyk 80]). Furthermore, in [Bonomi 83], it is shown that, under certain conditions, the number of necessary rearrangements is given by

$$P(j\,\text{rearrangements}) \approx e^{-(j-1)\mu}$$

where μ is equivalent to the "temperature" (cf. Section 6.4).

Thus, rearrangement appears to be an attractive technique with a low cost for the control system (in particular for time-division networks) and has a certain value in achieving zero or low blocking probabilities (it can be used on blocking structures or the number of rearrangements can be limited on a nonblocking structure).

However, it would still be necessary to extend the application to more general structures than the Clos three-stage networks and the systems derived in a "recursive" manner (cf. Section 6.3).

However, curiously enough, this technique appears only very rarely in effective implementations. Only in [Melas 83] has its use been taken into account in order to obtain nonblocking (probably for military purposes). Rearrangement has made it possible to reduce the size of the equipment by one-third. A small network was involved (3000 lines).

6.3 Complexity of switching networks

In the theory of complexity, large systems (i.e. systems formed from a large number of identical components and possessing a very large number of states) are studied from the viewpoint of the relation between their size and the service rendered. Attempts are made to answer the question of how many elements are necessary to accomplish a certain function. The theory finds an interesting application in the study of switching networks, the most "complex" systems that man has ever constructed.

Apart from speculative interest, this theory makes it possible to assess the cost-to-quality ratio of networks in operation and operational procedures in a very global manner.

Complexity is understood somewhat intuitively as the number of different situations which the switching network has to deal with.

Thus, let there be a switch (more exactly, a unit operating point-to-point connections) with N distinct inputs and N distinct outputs. This switch will handle the following situations.

- No input and no output connected (one situation).
- Any one input connected to any one output (N^2 situations).
- .....
- k inputs connected to k outputs; inputs and outputs can be chosen in $\binom{N}{k}$ ways and they can be combined in $k!$ ways.

Finally, our device (a very crude model of an actual exchange) will handle $F(N)$ distinct situations:

$$F(N) = \sum_{i \leqslant N} i! \binom{N}{i}^2$$

A concentrator which has N inputs and N outputs and where the input requests a connection to any of the idle outputs will see the following number of different structures:

$$F(N) = \sum_{i \leqslant N} \binom{N}{i} = 2^N$$

(same counting procedure). The "combinatorial" nature of the complexity is well expressed by these equations and shows an enormous increase in $F(N)$.

How many elements are needed to handle these $F(N)$ situations? We call the commutative capacity $C(N)$ and it is equal to the logarithm to the base 2 of $F(N)$. This is the minimum number of logical tests to be carried out in order to choose one of these situations. This will therefore be the minimum theoretical number of elements necessary to perform the switching function. However, it is important to note that this limit does not specify the nature of the "element".

As soon as N is sufficiently large, the Stirling formula gives the order of magnitude of the increase:

$$C(N) = \log_2 F(N) \approx N \log_2 N - 1.44\,N \tag{6.8}$$

for an "exchange" (defined as above) and

$$C(N) \approx N \tag{6.9}$$

for the concentrator [Bonomi 81].

The same limit is obtained for ordinary networks; if the basic element is the elementary switch with two states, a switching network with p of them will take 2^p different states some of which will perhaps be prohibited by the operating mode. This number must not be less than the number of situations to be handled:

$$\begin{aligned} 2^p &= F(N) \\ p_{\min} &\geqslant C(N) = \log_2 F(N) \end{aligned} \tag{6.10}$$

The above-mentioned limit due to Shannon shows that it could be possible to construct switching networks containing of the order of $N \log_2 N$ relays.

However, there is nothing to indicate how the limit in eqn (6.10) should be reached.

We consider here nonblocking structures; the network can always route all the calls. We shall examine the case of blocking structures at a later stage.

If we restrict ourselves to the examination of existing structures, the non-blocking Clos network (b $n \times r$ matrices at the first and third stages and r $b \times b$ matrices at the second stage with $r = 2n - 1$) possesses the following number of elements:

$$2bnr + rb^2 = (2n-1)b(b+2n)$$

Under the constraint $nb = N$, and with $n = b = N^{1/2}$, the number of elements is $6N^{3/2}$ (the minimum obtained for a value of n close to $N^{1/2}$ is also of this order). The order of magnitude of the increase, which is better than that of the square matrix N^2, remains above the limit given in eqn (6.10).

In fact, as soon as N increases ($N > 189$), the three-stage structure may be replaced by a five-stage network obtained recursively, the central stage of the three-stage network being replaced by a three-stage network and so on when N increases to seven, nine, eleven etc. stages. No analytical formula which gives the size of the optimum networks as a function of N is known. However, the increase in size remains greater than the Shannon limit.

Many researchers have made ingenious attempts to describe structures approaching the limit in eqn (6.10). Cantor has suggested a structure described by $4N(\log_2 N)^2$ [Cantor 71].

In all these studies the complexity of networks is expressed by the notation O (function of N); the multiplying factor has little importance, as it is only the function of N which guides the increase. The Cantor structure described by $4N(\log_2 N)^2$ is finally "better" than the three-stage Clos structure described by $4\sqrt{2}N^{3/2}$ even if $N = 16\,000$ (approximately), for the latter structure is expected to be more costly than the former structure.

Of course, in practice for the usual values of N (from 10^3 to 10^5), a factor which is too large would make a structure described by $O(N \log_2 N)$ prohibitive.

For example, Bassalygo and Pinsker have shown that there are actually structures $O(N \log_2 N)$. Unfortunately, their results (which moreover do not give the corresponding network) provide a limit of $67.7\ N \log_2 N + O(N)$, which is of little interest.

The graph in Fig. 6.4 adapted from [Marcus 77] illustrates the respective increases for different networks.

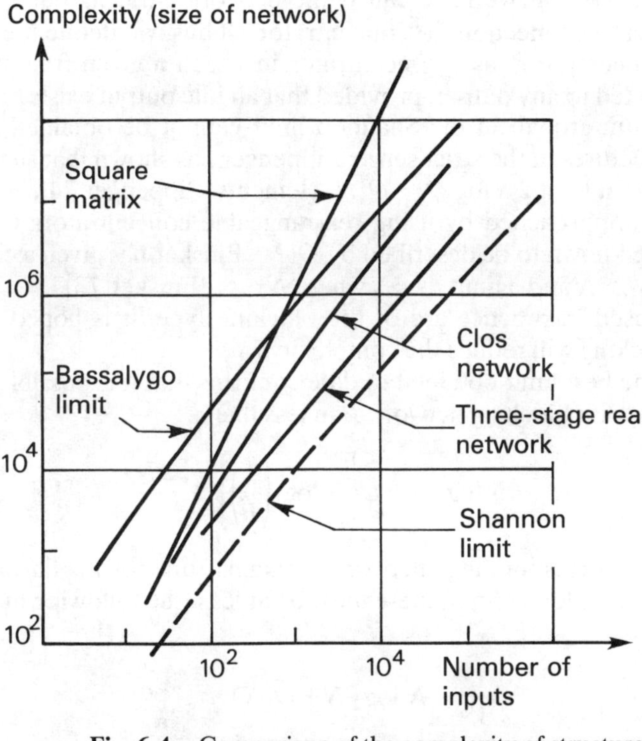

Fig. 6.4 Comparison of the complexity of structures.

The use of strictly nonblocking networks may seem a superfluous luxury; we could consider the comparative complexity of structures with a higher performance such as rearrangeable networks which are also nonblocking.

The rearrangeable Clos network ($r = n$) increases according to $O(N^{3/2})$. The application of the Clos system (with $r = n$) in an iterative manner, i.e. by replacing the central $b \times b$ matrix by a three-stage network and by applying the rearrangement strategy leads to an increase of the type $N \log_2 N$; more precisely, if the constituent matrices are all of the type 2×2, the (nonblocking) network will have $\log_2 N$ stages and a number of crosspoints equal to $4N \log_2 N - 2N$.

The use of 3×3 matrices slightly improves the limit: $3.8N \log_2 N - 3N$. The number of stages again increases as $\log_2 N$ and the complexity of the control, i.e. in this case the maximum number of rearrangements to be carried out on blocking, is in $O(N \log_2 N)$ [Pippenger 78].

It should be noted that for rearrangeable networks the complexity of the rearrangement must be taken into consideration. In fact, it has been shown in Section 6.2 that the cost of rearrangement in practice remains very far from the algebraic limit of the Clos network. It is certainly the same here.

Equation (6.9) gives a complexity described by $O(N)$ for the concentrator. It should be recalled that we have called the network carrying out "point-to-group" searches and connections a "concentrator". Thus, we define the "strictly non-blocking concentrator" as a concentrator in which a given free input can always be connected to any output, provided that an idle output exists.

The minimum growth in the Shannon limit cannot be obtained by non-blocking link structures in the strict sense; Pippenger has shown that such a concentrator requires at least $2N \log_2 N - O(N)$ elements [Pippenger 74].

The limit is approached by using rearrangeable concentrators for which the structures are known to be described by $O(N)$; Pinsker has given a construction described by $29N$ and a limit by $4N$ where $N \to \infty$ [Pinsker 73].

Networks used in practice are of the blocking type; it is hoped that the possibility of blocking will reduce the complexity.

In reality, the best limit obtained to date confirms that it is possible to construct a point-to-point blocking network B in less than

$$6N \log_2 N + O\left\{N \log_2\left(\frac{1}{B}\right)\right\}$$

As always, the corresponding network remains unknown. Furthermore, Pippenger has given a lower limit; there must be at least the following number of contacts:

$$3\frac{1+B}{1+A} N \log_3 N + O(N)$$

where A designates the offered traffic per idle input [Pippenger 75].

As can be seen, it is not clear that there is any interest in using networks with low blocking since the complexity limits of these networks may not be better than those of networks without blocking (rearrangeable in the strict sense). Nevertheless, it is possible that these results can be improved.

In any case, known structures which approach the limits, even if they turn out to be better in terms of contacts than the networks usually employed in telephony, do not seem to be directly usable because of the complete absence of modularity and the need for total rewiring when N increases and even for modifying the sizes of the matrices (the most striking case is the Clos network where the number of stages varies and where the size of the matrix is close to $N^{1/2}$).

However, they are suitable for applications where the network remains of the same size, i.e. it never evolves. This is the case for the "interconnection networks" used in data-processing systems containing a large number of processors operating in parallel (see Chapter 11).

We have used the words "crosspoints", "matrices" and the diagrams of link networks of the space-division type. In time-division technology the situation changes since the basic matrix which switches N incoming channels to N outgoing channels does not have N^2 contacts or even N^2 elements (integrated circuits or others); its "complexity" is of the order of N.

It therefore becomes possible in theory to ensure an increase according to $O(N)$ for a time-division network.

In fact, in this case it is difficult to speak simply of complexity; the equivalent diagram of the matrix remains an $N \times N$ matrix whose complexity is in N^2 but one of its dimensions is the time which is sampled at the speed of the multiplex. The buffer memories and the software whose influence on the complexity cannot be neglected are associated with this time dimension. The "complexity" of the time dimension is also a technological barrier beyond which it is impossible to work. In space-division technology the transition from N and $2N$ quadruples the equipment; in time-division technology the equipment needed doubles up to the limit of the technology where no more increase is possible. The cost is also related to the stage S which follows the matrix T and whose complexity is proportional to the "time dimension".

In general, it is difficult to define exactly the notion of cost linked to complexity. It combines the crosspoints in space-division or time-division technology, the directly associated supplies or peripherals but also the software, the associated data etc.

6.4 A thermodynamic theory

Since switching networks are complex systems, i.e. capable of being in a large number of states, it is tempting to apply to them the concepts and methods which physics has developed to study such systems, i.e. those of statistical thermodynamics.

The switching network is defined as a system evolving in a state space ("phase space"). A state e_x performs a given assignment between the inputs and the outputs.

E_x can be defined as the set of states performing the same assignment. We then search for the distribution of $P(E_x)$, the probability of finding a class of assignments.

The theory of statistical mechanics involves the "principle of entropy"; loosely, the accumulation of random microscopic changes causes (i.e. generates) a disordered distribution and an entropy:

$$H(E_x) = K \sum_x \{ P(E_x) \log_{10}(|X|!P(E_x))\}$$

measures this disorder and must be a maximum under the constraint of a given average traffic

$$A = \sum_x |X| P(E_x)$$

where $|X|$ indicates the cardinality of X (i.e. the number of calls which the state X connects).

As an example, let us show how the applications of this principle leads to interesting results concerning the rearrangeability. We call p_j the probability that the number of necessary rearrangements is equal to j $(0 \leqslant j \leqslant 2b-2)$;

$$m = \sum_j jP_j$$

designates the average number of rearrangements to be performed. The principle of entropy ensures that the observed distribution is that which maximizes "the entropy" defined by

$$S = -\sum_j P_j \log_{10} P_j$$

under the constraints

$$1 = \sum p_j$$

$$m = \sum jp_j$$

The solution of the problem is classical (see for example [Benes 65] and [Bonomi 83]). If b is sufficiently large

$$P_j = C\,\mathrm{e}^{-(j-1)\mu} \qquad j \geqslant 1$$

$$C = \frac{1}{m} = 1 - \mathrm{e}^{-\mu}$$

The simulation easily determines μ, e.g. $\mu = -\log_e(1 - p_1)$.

p_1 is the probability that a single displacement is necessary, i.e. that the central matrix accepting the displaced call does not route another request from the same incoming matrix somewhere else:

$$p_1 = 1 - \frac{nt}{r}$$

where t denotes the traffic per source; nt/r is the traffic per incoming link.

In [Bonomi 83], this approach is studied in depth (and a simulation confirms the relationship).

The same approach allows an elegant treatment of the waiting time calculation (see [Lutton 82]) for the waiting probability and distribution.

It is not possible to develop this theory any more here within the framework of this book; in order to do this, it would be necessary to start by introducing statistical mechanics. However, it does seem rather attractive. In particular it enables the concept of extensivity to be introduced; this concept is already known in thermodynamics and states that certain properties do not depend on the size of the system (the mass of a body is not an extensive variable in contrast with the temperature or the specific heat). The blocking probability under certain conditions is an extensive variable.

More details can be found in the references (and in particular in [Benes 65] and [Bonomi 83]).

References

This chapter is only intended to introduce the reader to a field of current research. If interested, this introduction can be completed by consulting the following references.

For link systems with delay, see [Altamirano 78], [Haugen 79] and [Kampe 77]; in [Hieber 70] an approach using equations of state is given.

For the complexity analysis, [Marcus 77] and [Masson 79] are valuable surveys and contain additional references.

The problem of path selection algorithms is probably the most important, especially in connection with multiple path selection (see chapter 11). See the studies of Ackroyd, Bazlen, Hofstetter and Melas.

[Benes 65] is the most important reference on rearrangements and thermodynamic theory and makes stimulating reading.

[Ackroyd 79] M. H. Ackroyd. Call repacking in connecting networks. *IEEE Transactions on Communications,* Vol. COM 27, No. 3, March 1979, pp. 589–591.

[Altamirano 78] M. Villen Altamirano. A calculation method for link systems with delay. *Electrical Communications,* Vol. 53, No. 3, 1978, pp. 217–224.

[Bazlen 73] D. Bazlen, G. Kampe, A. Lotze. On the influence of hunting mode and link wiring on the loss of link systems. *Proceedings of the 7th International Teletraffic Congress,* Stockholm, 1973.

[Benes 65] V. Benes. Mathematical theory of connecting networks and traffic. Academic Press, New York, 1965.

[Bonomi 80] E. Bonomi, M. R. Feix, G. Hebuterne. Information theory, structure and modeling of connecting networks (in French). *Annales des Télécommunications,* Vol. 35, No. 1, January 1980, pp. 39–59.

[Bonomi 81] E. Bonomi, M. R. Feix, J. L. Lutton. Telephone networks, statistical mechanics and non-random procedures. In *Numerical Methods in the Studies of Critical Phenomena* (eds J. Demongeot and B. Lacolle), Springer, New York, 1981.

[Bonomi 83] E. Bonomi, M. R. Feix, J. L. Lutton. 3-stage rearrangeable connecting network in the thermodynamic limit: number of needed reswitchings. *IEEE Transactions on Communications,* Vol. COM 31, No. 1, January 1983, pp. 155–157.

[Cantor 71] D. G. Cantor. On non-blocking switching networks. *Networks,* Vol. 1, No. 4, 1971, pp. 367–377.

[Haugen 77] R. B. Haugen. Erlang's ideal grading as a delay system. *Archiv für Elektronik und Übertragungstechnik,* Vol. 31, 1977, pp. 467–472.

[Haugen 79] R. B. Haugen. Link systems with delay. *IEEE Transactions on Communications,* Vol. COM 27, No. 9, September 1979, pp. 1261–1270.

[Hieber 70] L. J. Hieber. About multi stage link systems with queueing. *Proceedings of the 6th International Teletraffic Congress,* New York, 1970, Paper 233.

[Hofstetter 72] Von H. Hofstetter, E. Rokitta. The influence of path selection procedures on the traffic capacity of a PCM switching network. *Nachrichtentechnische Zeitschrift,* February 1972, pp. 92–101.

[Kampe 77] G. Kampe. Analysis of link systems with delay. *Proceedings of the International Conference on Communication,* Chicago, 1977, Paper 10.6.

[Jajszczyk 80] A. Jajszczyk, J. Rajski. The effect of choosing the switches for rearrangements in switching networks. *IEEE Transactions on Communications,* Vol. COM 28, No. 10, October, 1980, pp. 1832–1835.

[Lutton 82] J. L. Lutton. Information theory and statistical mechanics to solve a queueing problem in a switching network (in French). *Technical Report,* No. CNET/PAA/ATR/1199, February 1984, 38 Avenue de Gal Leclerc, Issy les Moulineaux, 92131 France.

[Marcus 77] M. J. Marcus. The theory of connecting networks and their complexity. *Proceedings of the IEEE,* Vol. 65, No. 9, September 1977, pp. 1263–1271.

[Masson 79] G. M. Masson, G. C. Gingher, S. Nakamura. A sampler of circuit switching networks. *Computer,* June 1979, pp. 32–48.

[Melas 83] C. M. Melas. Path rearranging in a data switching network. *IEEE Transactions on Communications,* Vol. COM 31, No. 1, January 1983, pp. 155–157.

[Pippenger 74] N. Pippenger. On the complexity of strictly non blocking concentration networks. *IEEE Transactions on Communications,* Vol. COM 22, No. 11, November, 1974, pp. 1890–1892.

[Pippenger 75] N. Pippenger. On crossbar switching networks. *IEEE Transactions on Communications,* Vol. COM 23, No. 6, June 1975, pp. 646–659.

[Pippenger 78] N. Pippenger. On rearrangeable and non blocking switching networks. *Journal of Computer and System Sciences,* Vol. 17, September, 1978, pp. 145–162.

[Pinsker 73] M. Pinsker. On the complexity of a concentrator. *Proceedings of the 7th International Teletraffic Congress,* Stockholm, 1973, Paper 318.

Chapter 7

STRUCTURE OF THE CONTROL SYSTEM

This chapter begins the second part of the book which concerns the study of the performances of exchange control systems. As with the switching networks, we start with a description of the elements involved, focusing on the traffic flow aspect.

There is less unity in control structures than can be found in switching networks. Many implementations which are extremely different have been carried out; SPC switching is still a very young and developing technology for solutions to have found any common ground.

We do not present all the possible implementations or even all those in existence. Our aim is rather to explain the functions and to define the elementary mechanisms as a help in analyzing present and future structures.

Two major concepts must be considered: the organization of the control system (monoprocessor or multiprocessor; centralized or distributed) and the organization of multiprograming on each processor.

Section 7.1 recalls the control functions (call processing, surveillance and maintenance). The general organization and the main categories of possible structures (centralized, decentralized and distributed) are also presented.

Section 7.2 describes the organization of the control system which is in fact a computer equipped with specific peripherals. Irrespective of whether it is centralized or distributed, it involves a certain number of elementary mechanisms which are summarized. It seemed useful to illustrate this description by summarizing a few implementations from the point of view of the architecture and the processing of the calls (see Appendix Chapter 7).

In Section 7.3 the operation of this system is explained, using the processing of a call seen from the viewpoint of the control system.

All performance studies show that the call-handling capacity is in the first instance governed by the capacity of the processors; Section 7.4 examines their operation in detail (allocation of time in particular). Finally, Section 7.5 concludes by summarizing and underlining the approach followed in this description, which may serve as a model for any evaluation study. In fact, such a study will always start with a description of the control system; it is necessary to extract this from complex and too detailed information.

This chapter will not deal with overloading and its control which is covered in Chapter 10.

A large number of aspects have been omitted; these do not relate directly to the problem of traffic flow (software, technologies etc.). Other works have dealt with these questions and it is worth referring to them (see references). We once more emphasize that this book is not a treatise on the art of constructing exchange control systems; we are simply attempting to understand the principles of these systems in terms of the aspects concerning traffic.

7.1 Functions and structure of the control system

7.1.1 The functions of the common control

The exchange's computer* is entrusted with an increasing number of tasks; because of this, considerable power is available, and this is exploited in order to relieve the operators of maintenance tasks, to monitor the proper functioning of the installation and to enhance the grade of service offered. Thus, the control system may be responsible for the following.

(a) Telephone tasks for which it is designed (e.g. call processing and billing). As will be shown in more detail in Section 7.3, this covers a large number of actions to be carried out in real time: surveillance of subscriber lines, reception of the dialing, their analysis ("translation"), path searches, signal transmission commands etc.

(b) Self-surveillance tasks. The computer with its peripherals is a complex system in which any fault is of great significance, implicating the integrity of the entire system and the service of all the subscribers. This leads to a duplication of vital units and the incorporation of test programs in the software aimed at detecting any fault very rapidly in order to eliminate the faulty unit.

(c) Maintenance tasks. In contrast with the preceding tasks, this involves supervision of the state of elements external to the control system (e.g. subscriber lines and circuits). These are less vital functions but they represent a considerable volume of work and have a large influence on the overall grade of service. Computers provide a powerful aid here.

(d) Technical management tasks (e.g. updating of the state of subscribers and of circuits, routing modifications etc. as well as traffic measurements).

* We recall that we have agreed for simplicity to use the term "computer" to designate the SPC system which controls the exchange irrespective of its structure.

Obviously, these activities have different weightings in the design of the system. Paradoxically, although tasks (a) are the reason why the computer is involved, tasks (c) and (d) make up most of the software. As far as tasks (b) are concerned, their urgency gives them priority over all the other activities. In a simplified manner, the comparative characteristics of the different activities are as follows.

- Call-processing activities: average amount of software (100–200 000 instructions); 70% of the "processor time" resource at the peak hour.
- Self-surveillance activities: very small amount of software and time consumed (for normal service); maximum priority.
- Management and maintenance: considerable amount of software (10^6 instructions); small amount of time consumed at the peak hour; normally low priority.

In the rest of this book, we shall not be concerned with self-surveillance tasks, the study of which is part of reliability theory; at the traffic level, these tasks are simply expressed in terms of a very small increase in the execution times of the programs and scheduling. Similarly, we shall not directly consider maintenance or management tasks. These activities normally have an urgency (and therefore a priority) which is lower than the processing of the call which they must not obstruct. However, care has to be taken that they are not indefinitely delayed; the processors must be in a position to allow them to be executed. This is particularly the case in "man–machine relations" which must always allow rapid intervention of the operators.

If we focus our attention on activities associated with the processing of telephone calls, it should be noted that, although these functions are almost identical in all implementations, the solutions adopted may differ considerably. Apart from the individual choices of each designer, the differences can be explained by technological evolution such as enhanced performances of the processors, reduction in their cost and software development. The central issue is that of the distribution of the functions: hardware or software; hard wired or programed.

Since the advent of the computer, the hard-wired logic of the crossbar system has been superseded by the "all-programed" solution which is extremely centralized (what we mean by this will be explained later); the cost of the processors and the need to make them profitable explain these choices. Since then, a compromise has been established, which leaves all tasks requiring common data and a high degree of intelligence to program-controlled logic; this is likely to evolve. The hardware (hard-wired logic) handles the repetitive and routine tasks; it only requires local data and is unlikely to be modified. There is thus a good compromise between the gain in speed of the hardware and the flexibility of the software.

Note that here "hardware" means the part of the system external to the stored program processors. This hardware may itself contain processors (cf. the "regional processors" of the AXE system).

7.1.2 The organization of the decision

The choice of control architecture results from a compromise between the costs of processors and software and the constraints on operation in real time; it therefore develops in step with advances in technology.

A distinction should be made between the concepts of a physical structure (the assembly of processors, memories and the switching system with strong intense coupling), the concepts of a logical structure (which governs the exchanges between the components, the linking of tasks and the allocation of resources and is the operating system) and the concepts of a functional structure (the division of the machine tasks into functional subsets).

The functional partitioning determines the installation of the functions (defined by the functional structure) in the physical structure.

The physical structure is a monoprocessor or multiprocessor.

The classification presented below is based on the type of organization of processors responsible for decisions and on the division of the operational control (logical organization).

Centralized architecture

The central processor which regulates the traffic flow is unique. It is involved in each phase of the calls and each call passes through it several times before it is terminated. The processor is the master of the system; it commands the periphery.

The principal drawback to this organization is the complete absence of modularity of the control; it is necessary to install all the computing power and even two computing systems right from the beginning in order to satisfy the availability constraints (see the next section).

Hardware costs are therefore not at all proportional to the carried traffic.

Another weakness of this organization is its incapacity to reach high performances which are made necessary by the development of the network and the high-level call-processing languages which facilitate the writing of programs to the detriment of execution times.

"METACONTA" (of ITT) is a typical example of this type of organization.

Decentralized architecture

This is a variation of the centralized solution which compensates in part for its weaknesses. The central processor, which is always present, delegates some of its tasks to slave processors, making the periphery modular.

The central processor maintains its scheduling role and remains an obligatory crossing point for the traffic. Again, this is the limit of its capacity. There is still little modularity.

The AXE system and the ESS4 system are examples of this type of architecture.

Distributed architecture

Several processors are involved in traffic handling and a call may possibly use several of them. In this organization the processors have lost their role as absolute masters of the process. They are rather seen as servers who are placed at the disposal of the calls via the periphery (and more particularly the subscriber units).

Such a structure has the advantage of being able to reach a large capacity with an excellent modularity. Handling parallel processing and ensuring synchronization are the difficulties here.

The distribution must be subject to evolution when technological development is borne in mind; we note that the term "distributed architecture" denotes quite different choices which we shall classify according to [Champeaux 79].

In a multiprocessor structure, it is important to define the "degree of privatization" of the resources, (in our case the processors). The regional processors of the AXE system are for the private use of the connected peripherals. In contrast, the central processor is of course common to all. In general, a processor will be common if all the peripherals can access it; it will be of the private type if a small part of the periphery uses it (these concepts make sense only if the reference system has been defined). The central problem of multiprocessor architectures is the intelligent handling of the degree of privatization. Extreme privatization simplifies the relationships between the elements; in contrast, the usual sharing which complicates the connections (which are necessary for complex addressing and contention problems) allows better use of the power and improved reliability.

A parallel aspect is that of specialization; the processors or more generally the resources may possess all the functions (universal distribution) or a small number of them (functional distribution). We should also mention the specialization of the memory or rather specialization of the data which it contains. The data are local (data associated with one unit) or common (the most numerous data involving the calls). In practice, common data are located in one or more common memories.

All these considerations lead to the definition of four main forms of organization shown in diagram form in Fig. 7.1.

- In the "universal distributed core" architecture; the control processors are seen by the periphery as a single unit; they function using dynamic traffic sharing.
- In the "functional distributed core" architecture, the processors are specialized, each function (translation, billing, path searches etc.) being entrusted to a group of processors (operating with load sharing for security). The E10 system is of this type.
- In the "dispersed" architecture, the core no longer exists (from the traffic viewpoint, however, a central management and maintenance unit may still exist). All the functions are provided by the periphery where each piece of equipment

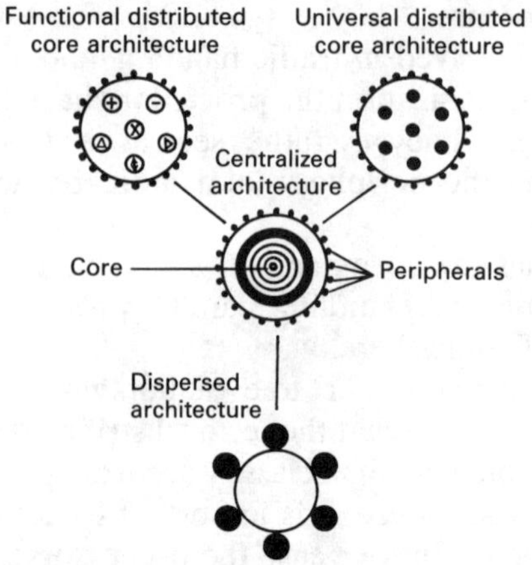

Fig. 7.1 The main forms of organization (according to [Champeaux 79]).

communicates with its neighbors. This is the extreme stage of distribution which probably ensures linear increase with increasing traffic.

These three structures are schematic and are never seen in their pure forms.

- The "decentralized" structure lies somewhere between the centralized structure and the dispersed structure.

7.1.3 Availability by redundancy

The permanence of service is one of the main criteria for the grade of service. Processors, memories, peripherals and connection units as in any data-processing system have nonzero failure rates and it is necessary to use redundancy to ensure an excellent reliability level (as an example, less than one total stoppage in 40 years in accordance with the French standards); each key unit is duplicated (at least). In the general case where n units are necessary for a function, $n+1$ (or even $n+2$) units will have to be installed and these will be omitted in the traffic calculations. The excess units are aimed at satisfying the availability constraint while the normal call-handling capacity must be guaranteed in the event of degraded operation.

Thus, in a centralized structure, two computing systems will be installed in order to satisfy the constraint of availability but a single system must be capable of carrying the traffic, observing the specifications of the standards. For this reason, evaluation studies are usually made in "simplex operation" (a single computing system in service as opposed to "duplex operation").

Several solutions can be envisaged for the organization of redundancy: load sharing, "microsynchronism" or standby. The last two solutions make sense only for active units (e.g. processors or a complete computer system).

Microsynchronism

The units work in parallel, receiving the same signals and producing the same responses, but only one communicates with the outside. A comparison circuit detects any disagreement, indicating a failure, and isolates the faulty element (the fault-finding mechanism which makes it possible to determine the fault is of extreme importance to the life of the system and must obviously have a high reliability).

This is the solution adopted, for example, for the central computer systems of the AXE and EWSD systems.

Standby

One processor (computer) is in operation; a second computer receives all the updating operations so that it can take over the handling of the traffic at any moment in the event of failure of the first. The calls in the course of being established at the time of the changeover are lost.

This is the simplest system to implement. It is used among others for the common control of system X.

Load sharing

In this solution the machines operate simultaneously and independently in the normal situation. We have a system with several servers (sometimes called a "pool" of servers). This is the solution currently used for the peripherals; this also applies to the universal core architecture of the preceding section. For active units, it is possible that they will have to inform each other of their actions (updating common tables).

It is necessary to resolve possible conflicts; a new call must only be detected by a single processor, and the common resources (switching network and equipment) can only be used by a single processor at any one time. Exclusion systems are therefore necessary. The central computers of the METACONTA which use this mode of operation proceed by interlacing the task cycles in order to deal with this problem (see Appendix to Chapter 7).

This is the most elaborate solution; it involves a channel conveying high-priority information (higher than that of the call-processing tasks) with efficient and reliable exclusion mechanisms.

This is also the most efficient; the dimensioning is usually made on n units but the traffic is most often carried on $n+1$ units. For centralized structures, this solution gives an overwhelming advantage to systems which adopt it; one computer system is necessary, and two are active. In fact, this does not double the capacity because of the exclusion phenomenon and the time spent in exchanging information; usually, a gain of 50% is achieved.

7.2 Constituent elements of the architecture

A control system can be regarded as being assembled from a construction set, the basic building blocks of which we present here. Of particular importance are the following.

- The connections between units for which a distinction is made between master–slave connections (processor–peripheral) and connections between processors.
- The organization of multiprograming (scheduling of tasks on the processor(s)).

7.2.1 General organization; peripheral units

The control system is built around processors which are similar or even identical with those of a classical data-processing system (at least in centralized systems); however, it will be provided with a set of specialized instructions (cable or microprogramed) such as bit-oriented operations, logic comparisons, processing of multiple words and interrupts. In fact, the reduction in the cost of "onboard" integrated circuits tends to promote the use of specialized processors (for the evolution of processors and their specific characteristics, see [Brilley 77]).

In order to collect the signals from outside (subscribers or circuits) and to act on them, specialized peripherals replace the traditional input–output units here.

These peripherals vary according to the architecture adopted. Thus, traditionally we find the following.

- Markers which are responsible for establishing paths in the connecting network. In space-division networks, these are relatively slow units positioning relays. In time-division networks, these are memories and control gates, which are more rapid units. These units are often responsible for testing the paths which they establish.

- The scanner. Subscribers and circuits are too numerous to be allowed to communicate individually with the central control system. The changes in state (unhooking, hanging up and dialing) are detected and stored in a peripheral and are periodically interrogated by a processor (a specialized processor of the subscriber line switching unit, an AXE regional processor, an ESS4 "signal processor" or even the central processor of METACONTA). Depending on the case, these peripherals will be more or less independent (see Section 7.3). In certain cases, the scanner is itself a complex processor (cf. the "signal processor" of the ESS4 system described in Appendix to Chapter 7).

- The units for sending tones to the users. On command, they send a signal to one end, subscriber or circuit.
- The signaling senders–receivers (communication with the neighboring exchanges). In channel-associated signaling, this involves simple units; they are complicated in the common channel-signaling system (see Chapter 11).
- Etc.

As well as these specialized units for telephone tasks, there are all the usual peripherals of a computing center.

- High-speed mass storage (disks and drums) used for the storage of data or telephone programs.
- Slower mass storage; magnetic tapes.
- Teletypes, consoles. The latter are useful for management activities and are usually considered as "man–machine interfaces".

The roles of these units and the operation of the exchange will be explained in Section 7.3 where we shall examine the process of a call.

The general operation of the system is that of a multiprogramed machine similar to a time-sharing installation; each call resides there, passing through the processors and the peripherals in a time of 10 to 30 s, but uses the units for a very brief time (say, 200 ms in total). Obviously, the processors cannot be in use all this time and multiprograming is the ideal solution which allows the best use of the resources but which gives each user the impression that he is in total control of the machine.

In such an organization the harmonious cohabitation of the programs is achieved by the job scheduler of the processor which schedules the sequence and priority in which jobs are to be run, i.e. allocates time according to the urgency of the activity and prevents any user from monopolizing the processor.

Here also several solutions have been put forward to solve this problem; Section 7.4 covers these.

7.2.2 Connection between processors and the periphery

This section concerns the connection between the central unit and the peripherals or, in a distributed system, between one processor and those of the peripherals which are assigned to it.

Peripherals are of two types: the usual data-processing units such as tapes, disk units or typewriters; the markers, scanners etc. which are specialized in telecommunication tasks. We refer to the latter as "telephone" peripherals for short.

Communication between the processor and classical peripherals takes place via "controllers" as in any mainframe system (the controller initiates and monitors the input–output operations of consoles, magnetic tapes etc.).

For telephone peripherals, the low volume of each individual transfer sometimes makes it necessary to economize the controller, the central unit carrying out the control itself; this is called, rather erroneously, a programed link.

The two solutions can be used concurrently (the controller for the data-processing peripherals, and the programed link for the telephone peripherals). The case of a specialized link to each peripheral can also be envisaged.

A distinction is made between two dialog modes. In both cases the processor controls the exchanges.

(1) Polling. The processor is entirely the master of the exchanges. A program of dialog with each peripheral is activated in cycles. The orders with the destination of the peripheral are placed in the buffer which is assigned to it; in each cycle, an order is sent via the bus. The response is recovered during a subsequent cyclic interrogation; simultaneously, a new order is sent if necessary.

(2) Interrupt control. Orders with the peripherals as the destination are sent to them directly or placed in a queue. At the end of the execution of an order, the peripheral interrupts the processor which reads the response, stores it for subsequent processing and sends the next order (which is at the head of the queue); exchanges are always made via the bus. The interrupt is preemptive (the central unit interrupts its task and saves its registers, in order to resume it after having processed the interrupt).

In terms of the time consumed by the control of the peripherals, comparison gives an advantage to polling with a heavy load; in this case, in fact, all cyclic activations are found to be efficient (i.e. they result in a transfer) and do not involve much loss of time in supervision. In contrast, interrupt control makes it necessary with a heavy load to carry out constant saving operations followed by reloading of the tasks in progress which involves an amount of management that can become quite considerable.

With a low load, polling loses its efficiency, most of the cyclic interrogations being useless (empty queues).

This relative efficiency of polling nevertheless assumes that the peripheral holding time remains constant and that the interrogation cycle has been well adapted (Fig. 7.2). Otherwise, an interrupt control is preferable, since it optimally adapts the relative speeds of peripheral units and of the central unit. Figure 7.2 illustrates the two modes of operation.

In these two modes, the processor itself controls the link and the corresponding queues (transmission of the information and orders, reading of the responses, saving in central memory and activation of the tasks generated by these responses); it may devote a nonnegligible part of its time for this and hence the interest in a solution using a buffer controller. In this case the processor only deposits its requests in the input queue which reads the corresponding data in central memory. The responses are placed there directly; the controller takes over the operations described above.

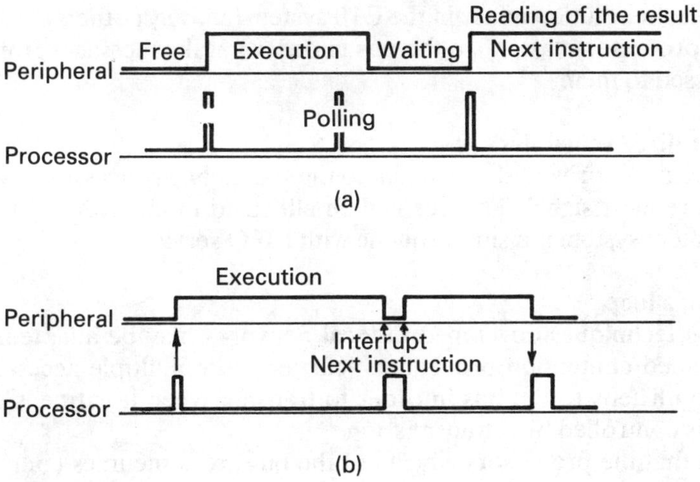

Fig. 7.2 Control of the peripherals: (a) cyclic control; (b) interrupt control.

Note that the peripherals involved are called "asynchronous"; after receiving the order, they become independent and so does the processor until the response is sent. Certain systems use "synchronous" high-speed peripherals. The scanner of the METACONTA system is an example of this as well as the signal processor of ESS4; the processor sends its order and stands by while waiting for the reply; the control unit can be used either to read or to write directly in order to receive or to transmit signals with a high time constraint, for example. There is simplification of the call processing at the cost of the performances.

7.2.3 Interprocessor linking

This point concerns more specifically distributed systems. Units with equal rank have to communicate. In contrast with the preceding case, this is not a "master–slave" link.

The case where the number of communicating units becomes large is excluded (cf. Chapter 11). A bus is sufficient to provide the exchanges.

It is necessary to deal with the problem of access to the bus. Several solutions are possible.

Polling

Each processor is allocated the right of using the bus in turn. The polling is carried out by a specialized scheduler which cyclically provides access for sending a single message or for a time slot. A distributed solution uses a token ring algorithm.

This is the solution in use in the E10 system (amongst others).

The processors "listen" to the bus in order to take messages from it which are addressed to them.

Centralized scheduler

Each processor is provided with a link to the scheduling processor to which they transmit a request signal. The bus is then allocated in the order of the arrivals (the equivalent system is a simple queue with FIFO service).

Use of a loop

The access techniques developed in local networks may be adapted: insertion loop, so-called contention procedures (carrier sense multiple access) in which each station listens to the bus in order to transmit when it is free. Collision of messages is controlled by retransmission.

Note that the processors which use the bus are sometimes completely different in nature; some are dedicated to the processing of calls, and others to control. The different constraints (holiday times) thus make it necessary to define access priorities. Care is also taken to break down long exchanges so that they do not disturb the overall operation even at low utilization rates.

7.3 Processing of a call

As an illustration, we shall consider an outgoing call established by a local exchange. Chapter 3 has defined the various phases in the establishment of the communication (preselection, selection and conversation). Here, we shall specify the role of the control computer.

Obviously, such a description depends on the technical solutions adopted; nevertheless, the diagrams which follow summarize the main characteristics which are found in all systems.

7.3.1 Centralized and decentralized systems

Preselection

When the subscriber unhooks, a switch or test point is positioned, which is responsible for signaling the event to the computer. The latter scans the lines in a cyclic manner; it periodically initiates the reading of a group of test points, detecting the changes in state which have occurred since the last cycle. In fully centralized systems the control unit entirely monitors the scanning. In a decentralized system a processor assigned to a group of lines monitors their state and transmits changes periodically or on request of the central processor (note the processing which relieves the central unit).

The computer will process the new call. The first operation consists of reading the subscriber parameters in memory (his state, his terminal type etc.). It is also necessary to find a memory zone ("register" by analogy with the electromechanical systems) to store the state of progress of the call.

Finally, it is necessary to find a dialing receiver unit (cf. Chaper 3). The computer knows their state and their address on the network. It searches for a path between the dialing receiver and the subscriber (in case of failure, it possibly changes the dialing receiver).

When the path is found, the memory is updated, i.e. the path (and the dialing receiver) is occupied so as to prohibit its use by another call; then it is physically established. The processor sends an order to the markers. Depending on their degree of intelligence, the order will be broken down or not; it is necessary to establish the path and to test it.

Having successfully completed its connection, the marker notifies the processor. This positions the dialing receiver which sends the dialling tone putting an end to the preselection.

Dialing; selection

The subscriber starts to dial. The successive digits are received by the dialing receiver. Depending on its degree of independence, this may serve as a simple electrical receiver, forcing the processor to check the pulses (the extreme case as in METACONTA), or it may read and store the whole number.

The successive digits accumulate in the register zone; at the end of a sufficient number of received digits, the computer initiates the selection of analysis of the direction ("translation") by looking up tables giving the address of the circuits for the prefix involved, search of a free circuit and of a path between this and the subscriber (still in memory) and search of equipment for transmitting signaling (depending on the type of signaling; for the common-channel signaling system, it is simply a matter of sending a message to the program or the separate unit which controls it).

Then, as previously, the memory map of the network is updated, and the connection is performed via the markers.

Thus the sending of the signaling starts with strong intervention by the control system and in parallel to the reception of the last digits.

During this phase the center transfers the necessary information for establishment (requested number, type of circuit, billing etc.) to the remote exchange; the transfer depends on the type of signaling (common-channel signaling system, multifrequency (MF) signaling or any other). In the usual case of MF signaling, a sender is involved. It may be independent or monitored by the control system. Similarly the responses from the remote exchange are processed in the sender or in the processor.

Transfer to ringing; conversation

The remote center responds and connects the circuit to the called party who unhooks. At the end of the signaling, the control system has released the equipment and the provisional paths. During the conversation, a memory area contains the necessary information for final billing and disconnection which will take place by means of the markers.

A complete description of the operation of the control system requires a similar scenario for all types of communication (outgoing and incoming) and for all possible outcomes (premature hanging-up, failures, occupation of the subscriber, fault etc.). The central processor intervenes at each step in order that the call can progress. This takes a path which is complex but determined by the type of system and leads from the processor to the peripherals and vice versa. In a decentralized-type system, returns to the central processor are less frequent.

7.3.2 Distributed systems

The communication process follows the same logical stages as previously with the essential difference that most of the tasks are received outside the central part of the control system; the "piloting" of the call most often involves the line switching unit (the terminal unit) which controls a group of subscribers or circuits.

The detection of the call is carried out as for a centralized system, the scanner unit being supervised by the line switching unit. Once the processor of this has detected the call, it completely takes over the preselection and most often the collection of the dialing which also becomes purely local.

The translation and the selection are carried out in the common part of the control system on request from the processor of the terminal unit. The control system carries out the translation, the path search and its establishment.

The difference between the centralized and distributed organizations from the viewpoint of the call process therefore mainly concerns the localization of the functions which are identical in both cases.

7.4 Multiprograming structure

Each processor (more generally each shared server) functions according to the multiprogramming" principle, each call occupies it for a few tens of milliseconds over a period of 10–30 s. Thus, several hundreds of calls may reside there simultaneously during the establishment of the call and it must share its time between them; therefore, after each elementary task, the call leaves the processor in order to occupy a peripheral, leaving it free to process another call which is waiting.

The processor has to carry out certain call-processing operations in a periodic manner with sometimes severe time constraints (generation or reading of signals in highly centralized systems).

Finally, the processor will often be responsible for various surveillance, maintenance and operation activities.

The relative urgencies of the different operations of the processor leads to careful ordering of the tasks. All systems do not allocate the same priorities, and the scheduling and timing mechanisms vary from one implementation to another. We shall describe some of them; these are job schedulers similar to those of any real-time data-processing system. They are classified as static or dynamic schedulers.

In general, we call a "task" the elementary unit of activity of the processor which can be interrupted or not. A task is completed by sending a message to the periphery or by placing it in a queue; similarly, each incoming message generates a task. A call consists of a finite sequence of tasks being executed in a given set of servers.

Static scheduling

There are two basic versions: cyclic scheduling and pure timed scheduling.

In the cyclic schedule the scheduler scans the buffers where the tasks are stored according to their type. Let us designate by *A, B, C,...* the different buffers. The task categories are identified by variable urgencies, where *A* is the buffer for urgent tasks, *B* for less urgent tasks, *C* for still less urgent tasks,.... The schedule can be carried out as in Fig. 7.3 which reproduces the schedule of system ESS1 [Hoover 70].

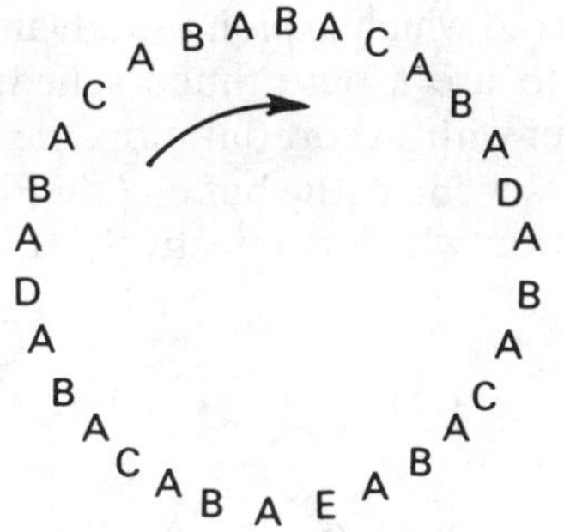

Fig. 7.3 Cyclic schedule for the ESS1.

The buffer *A* contains the most urgent tasks and is scanned most often. At each initiation, the processor examines the buffer. If it finds it empty, it passes to the next buffer; otherwise it empties it (following an order to be specified) before passing to the next buffer. A very common variant of this consists in only taking a finite number of tasks on each visit to the buffer. The buffer *E* contains the maintenance tasks with lowest priority. Its frequency of execution is an indicator of the load.

The advantage of this type of organization is the economy of supervision which it entails with heavy traffic (when the cycle lengthens, the relative time spent in supervision decreases). The tasks of accepting new calls are carried out less frequently, providing a natural load-regulation mechanism.

In the pure timed schedule a clock interrupts the processor at a regular frequency (e.g. once every 5 or 10 ms). The processor then executes the tasks of buffer *A*, passes to *B* until the last buffer where it possibly performs a loop. If the next clock pulse comes too early, it interrupts the cycle and the processor returns to *A* (see the example in Fig. 7.4).

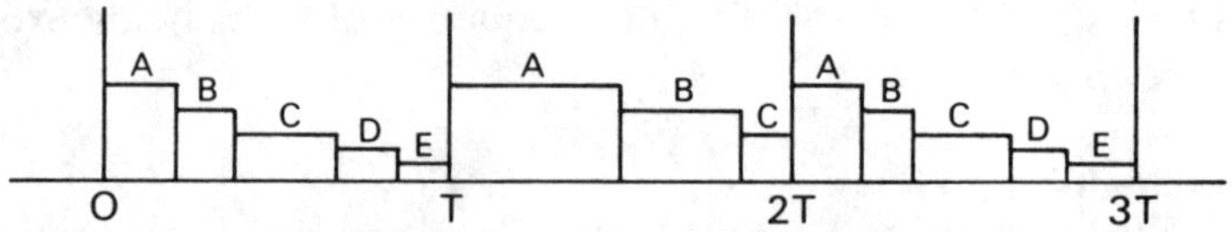

Fig. 7.4 Pure timed schedule.

When the load increases, the last buffers are no longer visited; the call-processing tasks which are kept there are not executed and there is blocking. The detection and regulation of overload are critical factors in this schedule. In contrast, the problems of precise measurement of time are easily solved.

The two types of scheduling are rarely used independently of each other; hybrid solutions are preferred which exploit the advantages of both of them.

Thus, it is possible to use a pure timed schedule for tasks occurring at precise regular intervals but with a short duration, the rest of the time being allocated according to a cyclic schedule (the buffer *E* in Fig. 7.4 contains the cycle in Fig. 7.3). This is the solution which is effectively adopted in the ESS1 system (Fig. 7.5).

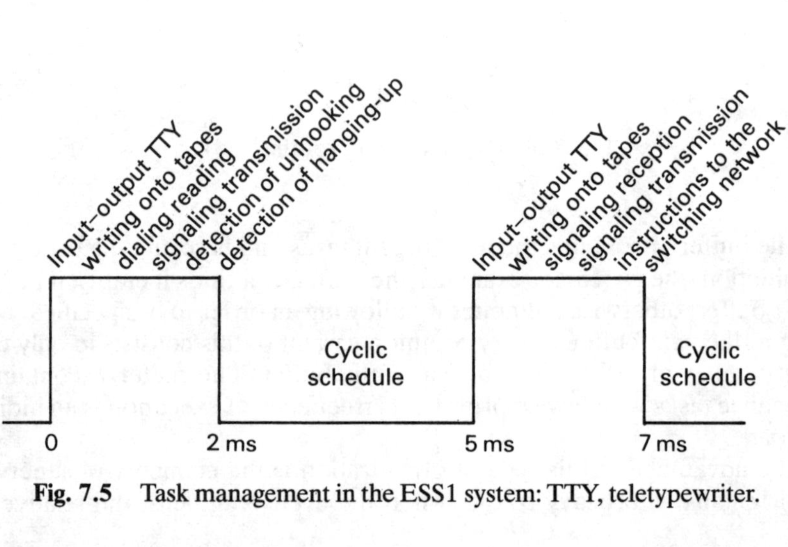

Fig. 7.5 Task management in the ESS1 system: TTY, teletypewriter.

Dynamic schedule

This involves using interrupt levels. In fact, all processors use interrupt levels for initialization, protection against faults and the inputs–outputs as has been described in Section 7.2. In addition, the dynamic schedule procedures use interrupts for call processing and not just to initialize the inputs–outputs.

Figure 7.6 gives an example; the interrupt structure of the central processor of the METACONTA system. A similar organization can be found in all real-time systems, the control systems of exchanges being an example.

In order of decreasing priority:

- * (0), * (1) Fault recovery
- * (2), * (3) IPL Communication between computers
- Telephone levels:
  - * (4) Clock H1 (10 ms)
  - * (5) Group marker
  - * (6) Line marker
  - * (7) Distributor
  - * (8) Exclusion between CU
  - * (9) Mass memory
  - * (10) Time delay (20 ms, 100 ms)
  - * (11) Level *P*
  - * (12) and following:
- Data-processing peripherals
- Base level (maintenance tasks controlled cyclically)

Fig. 7.6 Interrupt levels in the METACONTA system.

Levels 0–3 deal with serious events, implicating the integrity of the system and hence their absolute priority. Levels 4–11 deal with telephone task processing.

Here, the main level is 4; it is a periodic interrupt with absolute (preemptive) priority over lower levels which allows the signals to be exactly calibrated. During this interrupt the state of the lines and the dialing and signaling pulses are read. The control of this level is of the pure timed schedule type.

Level 11 ("level *P*") is where the call processing (analysis, path search etc.) is executed. Each level consists of a scheduler which links the tasks and distributes the resources and specific programs. It has a table of tasks to be carried out processed in the order of arrival.

The main telephone levels are assigned to the control of a class of peripherals. The priority between the different tasks is therefore regulated by interrupts.

In the preceding examples the priorities are provided by the interrupts. The interrupt is therefore most often preemptive. (The program being executed is stopped when the interrupt arrives, and its data and registers are stored. The interrupt program takes over the processor. On release, the interrupted program is resumed.) Software priorities can be used jointly which may complicate the rule.

Thus the central processor of the AXE system uses a priority level structure as shown in Fig. 7.7.

	* (0) HAL	
In	* (1) TCL Maintenance, surveillance	
	* (2) IPL	
order of	* (3) RPL1 Reception of signals coming from the regional processors RP	
	* (4) RPL2 Transmission of signals to the regional processors RP	
decreasing	* (5) THL1 Clock (10 ms): time measurement, intervals	
	* (6) THL2 Job buffer A, JBA	Telephone activity levels
priority	* (7) THL3 JBB	
	* (8) BAL1 Base level, JBC	
	* (9) BAL2 Base level, JBD	

Fig. 7.7 Task organization of the AXE processor.

The THL1 (traffic-handling level) and BAL1 (base level) are hardware levels and the THL2, THL3 and BAL2 are software levels. The rule is then as follows.

A program of level THL1, THL2 or THL3 has absolute priority (preemptive) on BAL. Within THL, the priority of THL1 over THL2 or of THL2 over THL3 is not preemptive (THL2 is perhaps waiting for the end of the task in progress of type THL3 before taking over). Similarly, BAL1 does not interrupt BAL2. The levels RPL and above interrupt THL and BAL.

Relative to cyclic control of the type in Fig. 7.3, there is true adjustment in real time; low-level tasks (which are nevertheless necessary for progress of the call) have full access to the central unit time at low load. At high load, urgent tasks are not held back by those of low priority. We meet again the fundamental dilemma of "cyclic control–interrupt control" which has already been mentioned in Section 7.2.1.

The following points should be noted.

(1) As in the pure timed schedule, the upper levels may suppress the base levels and load regulation is critical. This is especially true for centralized structures (see chapter 10).

(2) Priority in the AXE type of organization is generalized by introducing the concept of preemption distance. Let the levels be numbered from 1 to n in order of decreasing priority; the preemption distance of the level j will be the number δ such that j has nonpreemptive priority on $j+1,\dots j+\delta-1$ and may interrupt $j+\delta, j+\delta+1,\dots$. For the METACONTA system, $\delta=1$ uniformly; for AXE, $\delta=2$ for THL2, $\delta=1$ for THL3 and $\delta=3$ for THL1 (Fig. 7.8).

The properties of these schedules are analyzed in [Herzog 75].

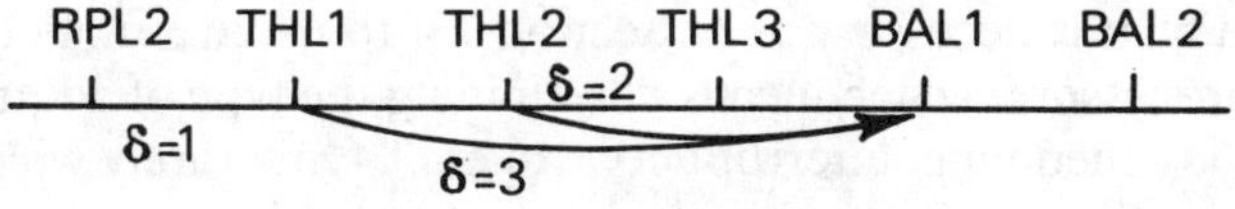

Fig. 7.8 Priority levels for AXE.

7.5 Conclusion; modeling of the control system

In this chapter the structure and operation of the control system (in at least those concerning the call processing and traffic flow) have been described. We shall now summarize this information.

- The performance of each function (the establishment or release of a call of a given type) represents a succession of operations similar to that described in Section 7.3. The elementary operations are handled by the processor(s) and the peripherals. Each call describes a path in the network of servers. The path is fixed for each type of call but varies considerably from one type to another.
- The processors form the operating nodes of this system; in particular, it is important to understand how schedulers distribute the processing resource between the calls actually present as a function of the urgency of the tasks and the load state.

In this chapter, as in the rest of the book, the attention given to the structure of the control system is dictated by the usage which is required. Chapters 8 and 9 are concerned with the evaluation of performances but it is necessary to start by constructing a model of the control system based on the understanding gained from this chapter.

In practice, modeling (setting up a model and theorization of the operation characteristics of the physical system) turns out to be an arduous and laborious enterprise. It is always an art but a plan of action can be indicated.

(a) The assumptions. Fault-free operation under normal conditions is assumed (without exception); serious overloading will be considered in Chapter 10. Nevertheless, a "simplex" state (e.g. a single computer in service in a centralized system) can be assumed according to usual principles.

(b) The first stage in modeling consists in drawing a diagram of the control architecture (types of peripherals, possibly decentralized processors, organization of the control system, connection to the periphery etc.) as was described in Section 7.2.

(c) The interesting types of call are then identified and the organization of the processing of these calls is constructed as in Section 7.3. In general, it is tedious and not rewarding to carry out this operation for all types of communication (see Chapter 8).

(d) Finally, it is necessary as in Section 7.4 to make a model of the active servers (the processors) which means determining the type of control of processor time (cyclic scheduling, interrupt control etc.). Only rarely will a single type of control procedure be dealt with; the control procedures used in practice are of the hybrid type. As will be seen in Chapter 9, this is the main difficulty in modeling activity; the existence of tasks under clock or cyclic control makes it difficult to apply the tools of queuing theory directly.

The approach described in this way may seem very general; the following chapters will explain and show that the control system is a network of queues, the elementary servers of which have been described here.

It is applied to all types of structure and to all stages of development, i.e. whether it is a completed system (final evaluation) or a system in the course of development in which the respective merits of the solutions are compared as a function of the traffic, hardware and constraints.

Appendix to Chapter 7

Description of the common control of a few systems

As an illustration of the concepts discussed in this chapter and in order to show the differences between effective implementations, we shall describe the structure of the control system of a few systems. For a general description, encompassing the switching network, the periphery, the software etc. the references should be consulted.

The METACONTA system

Various versions exist (METACONTA 10C, 11 A etc.). The switching network uses space-division technology. Two load-sharing computers provide the control with very centralized architecture.

The peripherals are essentially of three types.

- Markers which position the relays of the connecting network and test the electrical continuity of the paths.
- "Slow drivers" which position the relays in the equipment (junctions and auxiliary relays) for processing signals not subject to severe time constraints.
- The scanner which in fact consists of the line scanner which detects unhooking and hanging-up and the fast driver which scans the terminal points of the signaling auxiliary equipment and dialing receivers in a synchronous manner and transmits signals (tasks with strong time constraints).

In this system the central processor intervenes at all stages. The organization of the time assumes great importance. A periodic clock (clock H1) controls the operation. Every 10 ms, a cycle starts, with the scanner activated by the processor (scanning of a packet of 32 subscribers and then the dialing receiver and active auxiliary equipment for reading the pulses and finally activation of the signal transmitters). The processor is then dedicated to low-level tasks (level P) and processing of the peripherals.

The cycles of the two central units are out of phase by 5 ms in order to avoid conflicts.

Communication between the central unit and the periphery is carried out by means of interrupts using the algorithm in Section 7.2. The control of the processor time is also carried out by interrupts, i.e. the arrival of the interrupt from the periphery initiates the reading of the response, the sending of the next order and also the immediate activation of the corresponding call-processing task. The structure of the interrupt levels has been described in Section 7.4.

The AXE system

The AXE system developed by Ericsson exists in space-division and time-division switching versions. Contrary to the preceding system, it is largely decentralized; the regional processors relieve the central processor of repetitive tasks (reception of digits, detection of unhooking and hanging-up, scanning of test points etc.).

Each regional processor controls a set of equipment combined into extension modules (junctions, switching network etc.). The regional processors are duplicated for security and work by load sharing. The central processors function in microsynchronism.

Each regional processor is a computer with its processor, its own data store and programs. The regional processor's executive cyclically scans the extension module test points and calls in the corresponding programs. The cycle period is 5 ms.

The communication between the regional processor and the central processor is made by interrupts. Several buffers called job buffers serve to exchange the information. On the arrival of the interrupt RPL1, the central processor saves the message in the corresponding buffer (JBA and JBB for telephone tasks) and then resumes the interrupted activity. The message will be processed later according to its priority (see Section 7.4).

A duplicated bus RPB serves as the intercommunication medium.

In contrast with METACONTA, the central processor does not intervene directly in the routine tasks; the decentralization is very pronounced. However, the processor has to dedicate some of its time in communicating with its periphery (a large number of regional processors, up to 512).

The ESS4 system

This is a transit exchange with very high capacity (100 000 circuits and 50 000 erlangs) developed by Bell Laboratories. It uses a time-division switching network with a TS^3T structure.

The core of the system is the processor 1A specially developed for this activity. The architecture is of the decentralized type; the tasks of signaling and circuit scanning are entrusted to the signal processor (up to 24 per switch).

Each signal processor examines the test points which are entrusted to it (circuits up to 4096 per signal processor; MF senders–receivers, up to 32 per signal processor); each test point is scanned every 10 ms. If an event is detected, it is noted after confirmation in one of the four buffers of the signal processor (responses, new calls, dialing and abandonments).

The central processor and the signal processor operate at compatible speeds (the signal processors play the part of synchronous peripherals). In the central processor, two interrupt levels are dedicated to call processing: the base level and the priority "interject" level. The latter is a clock with a cycle of 10 ms. The base level allocates the time in quanta of 3 ms. At the end of each quantum, it tests whether a cycle of 10 ms has been completed, in which case it hands over to the interject level (Fig. 7.9).

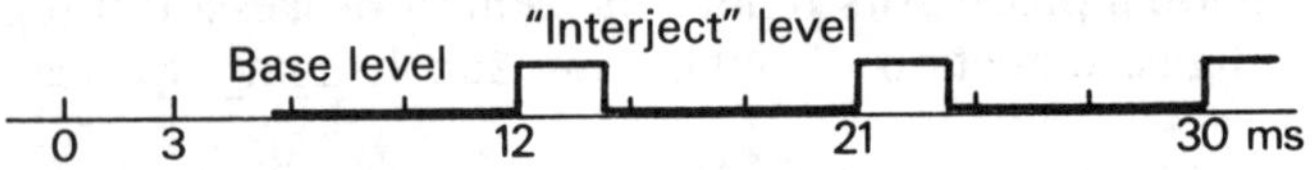

Fig. 7.9 Cyclic control of the ESS4 processor.

The processor directly reads the contents of the signal processor buffers (or writes to them). The compatibility of the speeds makes it pointless to use queues. The base level scans the job buffers and runs the corresponding call-processing programs. The interject level handles the tasks with more critical time constraints (the reading of the "signal processor response" buffers). The relative autonomy of the signal processor allows it not to use a preemptive discipline; the "planned interrupt" which the interject level accomplishes keeps all the flexibility of the interrupt control but avoids time losses due to saving operations.

The E10B system

In contrast with the preceding systems, the E10 system implements a true distributed structure of the "functional distributed core" type. We shall give a concise description of the E10B system [Fiche 81].

The control system is organized as shown in Fig. 7.10. The multiregister MR, the translator TR and the meter TX are specialized processors responsible for call processing. The maintenance unit OC is responsible for the system management (the link with the information processing center) and is not involved here.

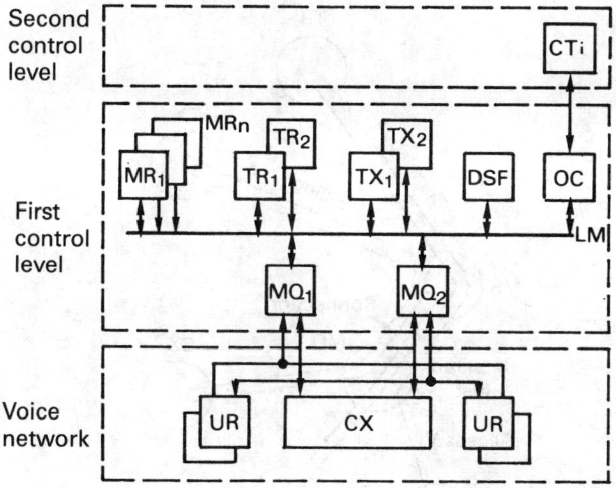

Fig. 7.10 The control system of the E10B system [Fiche 81]: MR, multiregister; TR, translator; TX, meter; MQ, marker; UR, line switching unit; LM, connection bus; CX, switching network; OC, maintenance unit.

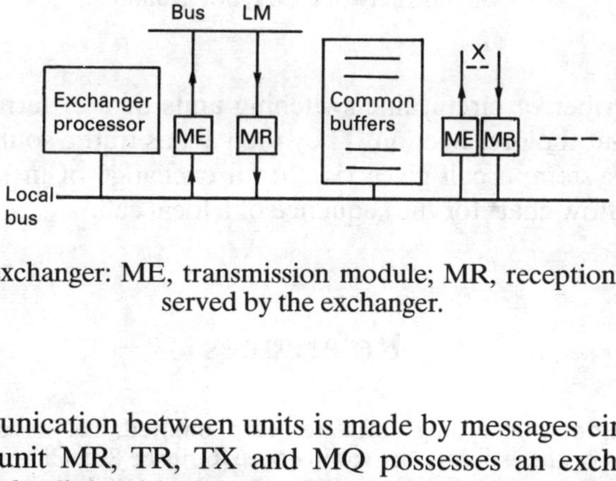

Fig. 7.11 The exchanger: ME, transmission module; MR, reception module; X, unit served by the exchanger.

The communication between units is made by messages circulating on the bus LM. Each unit MR, TR, TX and MQ possesses an exchanger which is responsible for the link with the bus. The exchanger is an independent unit organized as shown in Fig. 7.11.

The transmission module ME and the reception module MR are automatic devices responsible for the communication between the common buffer and the outside (the bus LM or units). The processor initializes the exchanges and controls the common space. The exchanger unloads the processors which it serves with communication tasks. Its essential role in terms of regulation will be seen in Chapter 10.

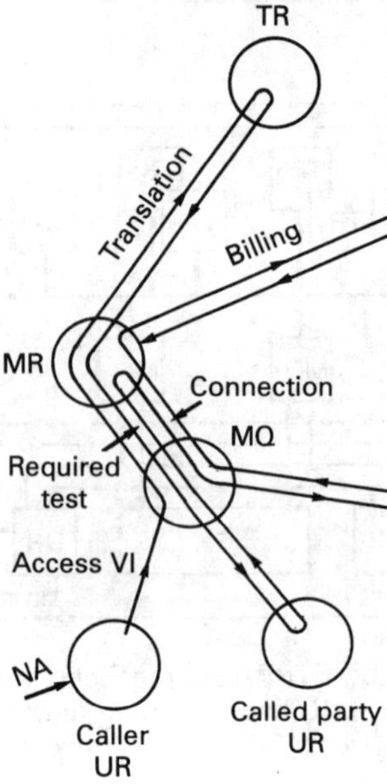

Fig. 7.12 Exchanges of messages produced by a local call in E10B. [Fiche 81]: MR, multiregister; TR, translator; TX, meter; MQ, marker; UR, line switching unit; CX, switching network; NA, not available.

The subscriber or circuit line switching units or user terminals detect the calls, carrying out the preselection. They then act as traffic sources with respect to the control system; a call gives rise to an exchange of messages. Fig. 7.12 reproduces the flow chart for the sequence of a local call.

References

Many considerations concerning software were omitted, i.e. its organization, data management etc. For these points, see for instance [Grinsec 83], Chapter 3, and a special issue of the *IEEE Transactions on Communications,* June 1982.

The history of electronic switching is related in a series of papers by Lucas (in French, unfortunately) [Lucas].

In [Brilley 77] the processors and their evolution in the telecommunication field are described.

Instead of giving an endless list of references, we urge the reader to refer to the *Proceedings of the International Switching Symposia,* especially the following: the symposium held in Paris, France, in 1979; the symposium held in Montreal, Canada, in 1981; the symposium held in Florence, Italy, in 1984. These contain descriptions of the main systems in use. Special issues of manufacturer reviews have been devoted to their systems (see [Hoover 67], [AXE 76], [BSTS 77] and [ITT1240 81]).

The description and specification of common control functions have benefited from the use of formal languages (see [Gerrand 82] and [Jallo 82]).

[AXE 76] *Ericsson Review, Special Issue,* No. 2, 1976.
[Brilley 77] B. E. Brilley, W. N. Toy. Telecommunication Processors. *Proceedings of the IEEE*, Vol. 65, No. 9, September 1977, pp. 1035–1313.
[BSTJ 77] ESS4 System. *Bell System Technical Journal, Special Issue,* Vol. 56, No. 7, September 1977.
[Champeaux 79] J. Champeaux, J. M. Pitie. Centralized–distributed control (in French). *Proceedings of the International Switching Symposium,* Paris, 1979.
[Fiche 81] G. Fiche. Methods for performance evaluation of the E10B system (in French). *Commutation et Transmission,* No. 4, 1981, pp. 69–76.
[Gerrand 82] P. Gerrand, E. Bierman. An overview of SDL, the CCITT specification and description language. *Telecommunication Journal,* Vol. 49, No. 5, 1982.
[Herzog 75] U. Herzog. Optimal scheduling strategies for real-time computers. *IBM Journal of Research and Development,* September, 1975.
[Hoover 67] E. S. Hoover, B. J. Eckhart. Automatic overload control for a stored program switching machine. *Proceedings of the 5th International Teletraffic Congress,* New York, 1967.
[ITT1240 81] ITT1240. Digital exchange. *Electrical Communication, Special Issue,* Vol. 56, No. 2–3, 1981.
[Jallo 82] W. D. Jallo, F. P. Coakley. The call processing multiparty software implementation *Computer Communication,* Vol. 5, No. 6, December 1982, pp. 291–297.
[Lucas] P. Lucas. Electronic switching in the world (in French) (a series of papers on evolution and features of systems). *Annales des Télécommunications,* Vol. 28, Nos. 5–6, 1973, pp. 201–243 (the first paper). *Commutation et Transmission,* No. 2, 1979 (the last paper).
[Grinsec 83] Grinsec. *Electronic Switching* (2 volumes), *Studies in Telecommunications,* Vol. 2, North-Holland, Amsterdam, 1983.

Chapter 8

EVALUATION OF THE PERFORMANCES OF THE CONTROL UNITS

In Chapters 8 and 9, we shall examine the methods for the evaluation of the performance of the control units. The importance of this subject makes it necessary to divide the presentation into two parts: this chapter serves as an introduction and discusses the "nonprobabilistic" approaches; probabilistic methods will be covered in Chapter 9.

The purpose of the evaluation is to verify conformity to the specifications (the standards mentioned in Chapter 1), to provide an aid in the design of the systems and to optimize their choice. Within the framework of the control structures, Section 8.1 defines the objectives which, as will be seen, are more varied than in the first part of this book.

We shall begin by constructing a very refined traffic model, defining the type and volume of the traffic as well as the user behavior. Section 8.2 contains this.

We then review the different methods which can be used. Proceeding from the simplest to the most complex (in the sense of intelligibility if not application), we discuss environmental simulation, an "experimental" method, in Section 8.3; then in Section 8.4 we deal with a deterministic method which is the simplest form of analytical method using a linear approximation.

8.1 Objectives of the evaluation

As for all systems which operate in real time, there is no doubt that it is important to know the capabilities of SPC exchanges installed in the network; however, evaluation has to start right from the beginning of design.

The study of the control system turns out to be difficult, first of all in terms of the results to be expected. Naturally it is best to evaluate the nominal processing capacity of the processor(s). This nominal capacity must first be carefully defined; in principle, the imposed standards specify the traffic flow conditions and the performances to be satisfied. As far as the control unit is concerned, the capacity is defined in the standards by response times such as "dial-tone delay", i.e. preselection delay etc.

This value of the capacity only makes sense relative to a given composition of the traffic (see Section 8.2).

The nominal capacity defines the number of busy hour call attempts; traffic above this figure must cause the violation of at least one performance criterion.

In the centralized type of control the availability constraints make it necessary to install two computer systems. The capacity, however, will be specified either in the "normal" or in a "degraded" mode, here in the Simplex configuration (only one computer active). The distinction only makes sense for a load-sharing operation. This point provides the immense attraction of this operating mode since, under normal conditions, two computing systems will be effectively available.

In a distributed control system the same constraints will specify a degraded mode (e.g. $N-1$ processors active out of N processors installed).

The nominal capacity may be derived from an overload specification if this represents the more stringent constraint. However, this case is distinguished from the objective of overload behavior. What would happen when there is serious deterioration in the offered traffic? The standards generally specify a "predictable" overload service grade (usually a 20–40% increase in the traffic), but the control system must be able to withstand higher loads by avoiding the breakdown mentioned in Chapter 10. However, here it is a matter of studying the behavior under the transient conditions of the monitoring mechanisms (which in principle are not involved in the overload of the standards). The subscriber behavior is of special importance in this study and we leave this until Chapter 10.

The dimensioning of the equipment (dimensioning of the peripherals, the junctions and termination equipment and also dimensioning of the queues and internal memories) is of course one of the objectives of the evaluation. In general, this is a complex problem in queuing theory because of the overlapping of the flow and queuing mechanisms. The case of the peripherals is usually handled by isolating them from the control system. This then reduces to the elementary teletraffic problem (see Chapter 2).

We emphasize the importance of correct dimensioning; as will be seen, capacity studies often assume a fluid flow at the level of these units. Thus, poor dimensioning introduces undesirable elements.

Optimization and choice of strategies are also important. Since the control system must handle traffic, it seems normal to use traffic flow criteria for making the right choices of structure and organization of hardware or software as early as possible in the design phase.

Here, it is not a matter of direct optimization but of comparison under identical traffic conditions. In the first instance the examination will be aimed at the identification of bottlenecks but also for example at the organization of the processors (cyclic or interrupt control; choice of job priorities; in general, organization of the multiprograming), at sharing and localization of the data in memory, at communications etc. The operations to be carried out may be made on the hardware (modification of scheduling; location of the data, transition to hard-wired logic as a last resort) or on the software (optimization of programing; writing in assembler; use of "flow analysis" which makes it possible to optimize the time consumed by a program by analyzing how much of it is taken up by the users there) (see [Potter 73]).

The evaluation methods will not be expected to indicate these improvements. However, they can quantify their impact on the nominal capacity. The estimated gain will allow the designer to decide whether to carry out additional work or not.

We have reviewed the fixed objectives of evaluation; we shall now discuss the methods which can be used. Direct experimentation is excluded from these methods; as has been already mentioned in Chapter 1, evaluation should be carried out during the design stage or before full loading which occurs a long time after the exchange has started.

The techniques which we are going to describe require data acquisition (data internal to the units to be evaluated and data external with respect to the subscribers). This is the traffic model which will be developed in the next section.

8.2 The traffic model

The processors handle several types of communication: incoming or outgoing calls but also attempts which fail because of incomplete dialing, rejection in the network or fact that the subscriber is busy. Although these failures do not contribute to the carried traffic nor to revenue, they do burden the active resources of the control system (and the processor time in particular). The evaluation must therefore take into account, in particular as they partly originate outside the exchange; it would not make much sense in guaranteeing a number of carried calls which did not reflect the activity of the control units and which changed with the evolution of the outside network. For this reason the capacity is specified in terms of busy hour call attempts, the traffic model relating these attempts to the effective traffic.

It should be noted that the traffic handled by the switching network is expressed in erlangs and only took established calls into account which simplified its description. Here, a distinction is made between the types of call and their degree of completion, the processors intervening more or less according to the type.

First the distribution of the appearance of the calls is defined. As we saw in Chapter 2, it is customary to consider Poisson arrivals, this assumption being justified by the large number of independent sources (Chapter 2, Theorem 2). All measurements carried out to date have confirmed this distribution although very few have in fact been made.

The composition of the traffic should then be defined. The breakdown of traffic is as follows.

- Outgoing calls.
- Local calls.
- Incoming calls.

The breakdown varies with the location (5% of local traffic in local exchanges of large cities, and 20–30% in a rural area).

For outgoing traffic, the percentages are specified for each direction or rather for each class of outgoing direction defined by the number of digits dialed.

Most of the time, a distinction will be made between the following.

- Local calls (to the same area) which require typically 6–8 digits to send.
- Calls to the trunk network which need the area code plus the local number, say 8–11 digits.
- International automatic calls which consist of the special access code to the international direct distance dialing, then the country code and finally the called number which means that usually more than 12 digits have to be sent.
- Special services (emergency codes and calls to the operators) which require 2–3 digits.

Of course, these distinctions are important because the dialing processing time may depend on them. In addition, the dialing time and therefore the occupation of the recording memories depend directly on them.

The percentages of calls to the trunk or international networks vary enormously depending on the location.

In contrast, the behavior of the subscriber during dialing will have a more universal character provided that a distinction is made between the users equipped with dial-up terminals or those with keypad terminals. Table 8.1 gives a few typical values taken from [Dietrich 75].

Table 8.1

	Average time (s) *for the following dialing systems*	
	Dial	*Keypad*
Waiting before dialing	3	2.5
A pulse train	0.55	0.21
Between successive pulses	1.3	0.5

Finally, of course, it is necessary to specify the fate of the call. For an outgoing call, this may be terminated by the following.

- The absence of dialing (faulty operation nevertheless occupies the processor).
- Incomplete dialing (hanging up after the first, second, ..., digit).
- Blocking in the exchange (no outgoing junctor) or in the network (because of a fault, saturation of the network etc.).
- The called subscriber is busy.
- No reply from the called subscriber.
- Sometimes, by a conversation.

Here, also, the proportions of each category strongly depend on the exchange. Thus the fact that the called subscriber is busy will be more frequent in an exchange handling "professional subscribers", where in contrast the absence of reply will be rarer than in an exchange serving "residential subscribers" (rarely busy but often absent). Similarly, the behavior of the callers will vary.

Nevertheless, as a guide, we shall give a few typical values in Table 8.2.

Table 8.2

	Breakdown (%) of outgoing calls	*Breakdown (%) of incoming calls*
Call without dialing	10	—
Incomplete dialing	5	3
Blocking in the network	5	—
Called subscriber busy	15	15
Called subscriber absent	10	10
Conversation	55	72

The network "filters" some of the ineffective calls which explains why the percentage of success is higher for incoming than for outgoing calls.

These values also vary with the load level (e.g. increase in the blocking in the network).

In Table 8.2, we have neglected failure inside the exchange (no dialing equipment or no path in the switching network). In fact, if it is correctly dimensioned, these circumstances will remain rare (the probability of these events is of the order of 5×10^{-3}–10^{-4}).

Table 8.3 specifies the behavior of the user for these different events.

Other more detailed values concerning the subscriber behavior are given and discussed in the references.

If necessary, this description is completed by giving the telephone environment of the center (type of signaling link with neighboring exchanges).

Table 8.3

Time of listening to the busy tone (the called subscriber is busy or congestion)	5 s
Ringing time (before unhooking)	12 s
Ringing time (no reply)	20 s
Conversation time	2–5 min

We draw the following conclusions from this section.

- The need for a detailed traffic model.
- The difficulty in constructing it.

This difficulty is due to a large number of parameters: the cost of reliable measurements and also the variations in these parameters for subscriber behavior with country, even the region (urban or rural area), the traffic level, the hour of the day (in particular with respect to the duration of the conversation) etc.

The set of values which forms the traffic model is often called the call mix or call type distribution.

In practice, such a complete model is rarely used except perhaps for environmental simulation. Similarly, it is not systematically adapted to each exchange; the data acquisition which is necessary for the model involves considerable effort.

8.3 Environmental simulation

Direct measurement on an actual machine obviously seems to be the most reliable method of evaluation. We have already mentioned the difficulties of making such measurements, e.g. the need to have available a complete installation loaded to maximum capacity, the difficulty in interpreting the results, nonreproducibility and the cost of a set of measurements.

Environmental simulation attempts to overcome these drawbacks. This method was originally developed for debugging computer programs (first reported in 1967). The large quantity of programs (sometimes 500 000 to 1 000 000 instructions), together with the difficulty in debugging in real time where it is necessary to correct software errors and hardware faults simultaneously (and also the possibly resulting disturbance for subscribers), made it necessary very early on to find efficient methods of debugging the software on a model.

The principle of environmental simulation is very simple; only the control unit and its programing need to be tested. The "environment" is therefore removed, most often at the level of the periphery (the simulated part which is

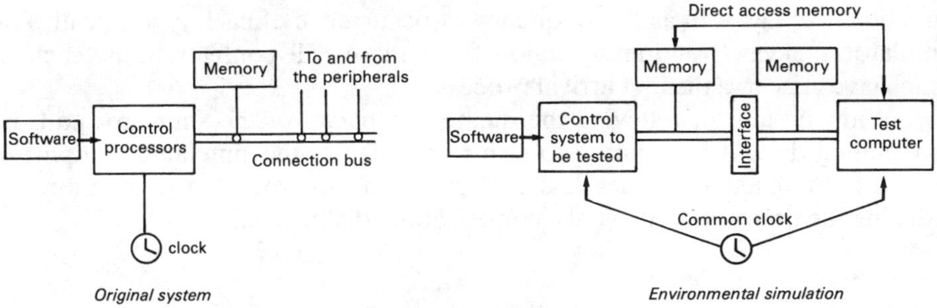

Fig. 8.1 Environmental simulation.

kept depends on how advanced the implementation is); the periphery and the external environment are replaced by a computer simulating the responses (Fig. 8.1).

For applications for debugging programs, it is worth synchronizing the clocks of the two machines in order to obtain a completely controllable and reproducible operation. It is also advantageous if it is possible to suspend the activity of each of the computers while the other is working. In fact, this enables the time to be perfectly controlled (with the aim of always being able to reproduce the sequence of experiments exactly, an indispensable condition for any validation and detection of faults). This also leaves the simulator the time to carry out all desired verifications (direct access is made to the memory of the computer to be tested).

How does an environmental simulator work? It is copied from a classical simulation program. The simulation is activated by the following.

- Orders from the control system under test.
- Future events which it stores. As in a simulation, a table of future events (the response to an order, the arrival of a new call or another external event) is maintained. Each activation feeds this table.
- If necessary, the load can be completed by artificial traffic generators if the configuration allows this.

Such a simulator, which is designed for validation, is easily adapted to quantitative evaluation. The required measurement subprograms will be added there, together with the exact statistical description of the traffic for the generation of different types of calls (cf. Section 8.2); such a statistical description is not useful for validation.

An elegant method for generating external events consists in defining "call configurations" stored in memory; each element describes the type of call, e.g. outgoing or incoming call, and effective or ineffective. The configuration describes the set of interventions of the periphery for each call, the reactions of the users incorporating random events (intervals between digits, requested directions etc.) as well as the frequency of occurrence of each type of call. The simulator makes a random selection from these call configurations, each of which gives rise to a distinct arrival process.

With the aim of relieving the memory capacity of the simulator and the programing data, it is common to reduce the size of the simulated equipment; they are simulated ten times less and are made to work ten times more by reducing the time between events (conversation, dialing etc.).

8.3.1 Advantages and disadvantages of this method

Environmental simulation turns out to be an indispensable tool for developing systems which are already structured and in an advanced stage of development; the software must be available. The weaknesses or disadvantages of the method can be listed as follows.

(1) It is necessary to have available complete programs for call processing; this implies a sufficiently advanced design phase and the impossibility of returning to detected faults involving more than simple programing details. Similarly, the method does not make it possible to choose between options which the designers could not agree on because it was necessary to program them completely (operation of prohibitive cost). In summary, it is not possible to use environmental simulation during the design stage.

(2) It has been seen that it is necessary to "organize" the original traffic model in order to reduce the number of sources; what is the validity of such manipulations?

(3) It is necessary to construct a simulator and to program it; the debugging of the switching system must be accompanied by the debugging of the simulator. In addition, the development of the programs, the acquisition of data on the types of communication, fault etc. represent a considerable amount of work which can only be partly reused.

(4) There are many problems in reusing a simulator, e.g. the call configurations, and the fact that the simulated peripherals depend considerably on the machine. This influences the cost of the evaluation by environmental simulation. This is also determined by the cost of one experiment which is determined by the time; the ratio of the experiment time to the simulated time is about 10–30.

(5) On the positive side, it should be noted that the method measures correctly and easily the frequency of occurrence of the programs and the time that they consume (path search etc.). This is a very valuable asset, making it possible, for example, to optimize the programing of repetitive sequences (cf. the flow analysis in Section 8.1).

(6) It also makes it possible to analyze programs for controlling overload or breakdown in the laboratory.

8.3.2 Conclusion

Environmental simulation seems well suited to the development of programs and to the development of different versions which can be adapted to customer's requirements. On the basis of this application, and only if it is justified, it is possible without too much effort to derive a quantitative evaluation product which makes it possible to refine the values given by other methods exploited earlier in the development.

Having said this, it is an indispensable tool for developing certain programs, e.g. fault diagnostics and protection diagnostics. It is also a good method for testing and developing programs for controlling overload (cf. references).

8.4 A linear model

The deterministic approach is based on the assumption that the load of the shared units varies linearly with the traffic.

To be more exact, in a centralized system, the peripheral resources may always be adjusted, and the bottleneck for a correctly dimensioned exchange will be located in the central unit whose rate of occupation will vary linearly with the load.

In a distributed system, each active unit (processors, memories and switching system) will follow a linear distribution on its own account. We shall find that, in this system, groups of units operate with load sharing; the rate of occupation per unit will satisfy a linear distribution. The bottleneck will then result in the most loaded unit.

For simplicity and without loss of generality, we shall assume that the bottleneck is caused by a processor. If we then examine the load on this processor, it can be separated into the following.

- A load independent of traffic and location; the proportion of time is dedicated to scheduling tasks, controlling the monitor, various obligatory maintenance or surveillance tasks etc.
- A load independent of traffic but depending on location, more exactly on the amount of peripheral equipment, e.g. the amount of time dedicated to systematic scanning of the subscriber lines and the circuits which are under the control of the processor.
- A load varying with instantaneous offered traffic; each call requires an involvement of the processor. As a first approximation, it can then be assumed that this load is proportional to the traffic.

- The processor dedicates the rest of its time to nonurgent and therefore nonpriority background jobs. These activities of maintenance and updating do not concern call processing. However, they still cannot be delayed indefinitely.

The breakdown of the load given here is very general. However, an exact breakdown of the jobs into the above-mentioned categories can only be done by analyzing the operation of the processor in detail.

Figure 8.2 illustrates the evolution of the activity rate of the processor studied with the offered load.

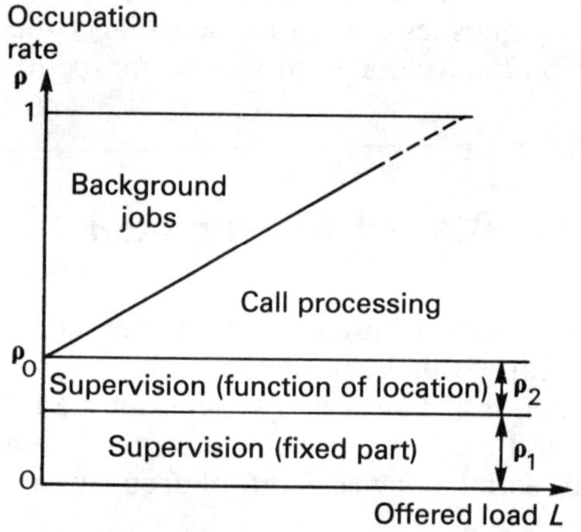

Fig. 8.2 Linear model: occupation of the processor.

We call L the load offered to the control system (the number of calls to be handled per unit time) and ρ the occupation rate.

The linear model constitutes a first approximation. In particular, it is only valid in a limited area; the assumption of linear growth is reliably justified in the neighborhood of a zero load. When the load increases, various phenomena can change the gradient; the overload detection programs will cancel certain sequences of the call processing (e.g. tests). The waiting and contention phenomena (increase in the number of interrupts and context-saving operations etc.) with respect to the availability of resources will produce an increase in processor time consumed per call; similarly, there will be an increase in the duration of path-searching programs in the switching network.

Once the assumption of linearity is accepted, another assumption is made relating the grade of service constraints to the load of the processor considered. Yet again, this assumption implicitly assumes that this processor produces the bottleneck responsible in the first instance for the degradation of service when the traffic increases.

Thus, since a response time (say a preselection time) is involved, the main cause of additional waiting relative to a low load lies in the access to the resources and mainly to the most loaded resource amongst them; it can be seen that there is a relationship between the response time and the occupation rate of the most loaded unit. The definition of a preselection time constraint therefore implies a maximum load and via the linear assumption the determination of the capacity L_{max} (the number of permissible call attempts per unit time). Since the definition of the permissible capacity must take into account any overload standards, it is necessary to assume that the linear model is still valid under overload conditions. Care should be taken that this value, denoted ρ_{max}, is a function of location (see below).

The application of the model requires a more exact knowledge of the operating mode of the processor. In particular, a distinction is made between cyclic control and interrupt control (cf. Chapter 7).

Let us consider interrupt control. First, we plot the straight line $\rho = f(L)$. We therefore set (Fig. 8.2)

$$\rho = \rho_0 + \bar{t}L \tag{8.1}$$

where ρ_0 is the no-load occupation rate and ρ the occupation for the load level L (in calls per unit time). Of course, the occupation rates relate solely to the load of telephone tasks; they represent the percentage of time which the processor spends in the telephone activity levels (in the absence of calls, the processor devotes its time to various background tasks so that its overall occupation rate is always 100%).

The parameter $\bar{t}$ in eqn (8.1) is the average time devoted to the processing of a call by the processor (which is of course expressed in the same units as L). This time is a function of the composition of the traffic handled by the center (the call mix).

ρ_0 is the sum of ρ_1 and ρ_2 where ρ_1 denotes the fixed part of the load and ρ_2 the part of the load which is a function of location. In practice, if the number of subscribers and the number of circuits controlled by the processor are denoted by N and J respectively,

$$\rho_2 = \alpha N + \beta J$$

It is necessary to evaluate the parameters $\bar{t}$ and ρ_0 (or better still $\bar{t}$, ρ_1 and ρ_2). Several approaches are possible.

- If the speed of the processor is known, the instructions of the call-processing programs can be counted, distinguishing between sequences belonging to $\bar{t}$, ρ_1 and ρ_2.
- If a model is constructed, ρ_0 can be measured with it and maybe even ρ_1 and ρ_2 by making the parameters N and J vary. The measurement is simple to carry out since the processor with no load spends the proportion $1 - \rho_0$ of its time at the background level, provided that it has installations of typical size available, however.

• For the evaluation of $\bar{t}$, it is necessary to return to the composition of the traffic. Let f_i, $i=1, 2, \ldots, n$, be the proportions of the different types of call (outgoing, local, incoming and various destinations), t_i the processor time is defined for a call of type i and

$$\bar{t}=\sum_i t_i f_i \tag{8.2}$$

The t_i account for the total time taken up by the call (all interrupt levels merged). They may be evaluated by counting the instructions or from the model (which is an imprecise method which takes into account the case when the load level is unavoidably low).

• If a site exists which carries actual traffic, it is possible to measure ρ_0, at nighttime for example, and $\bar{t}$. In this case, f_i cannot be controlled of course. One solution is to measure $\bar{t}$ at the same time as the f_i values and to derive the t_i values from them; measurements are taken at different times (day, night etc.) in order to have different samples of f_i.

It thus remains to determine $\rho_{\max}$. The maximum permissible capacity is derived from eqn (8.1).

$$L_{\max}=\frac{\rho_{\max}-\rho_0}{t}$$

Although this is a very simple method, it is capable of answering a large number of questions, and in particular what happens when the center evolves?

First, it is necessary to solve the problem of $\rho_{\max}$; how does it vary with $\bar{t}$ and ρ_0?

Here, we are faced with a queuing problem which cannot be handled without specifying the operating mode. We shall adopt only a simplified approach.

We represent the control system by a single queue. The response time is a function of only the overall load and $\rho_{\max}$ is constant (of the order of 0.7; for a queuing system the response time increases considerably beyond this value).

Other empirical arguments can be found in the literature; the waiting time is a function of the time actually available for the call processing. Here, because of the fixed load ρ_0, this time is only $1-\rho_0$. This means that the equivalent service time is no longer $\bar{t}$ but $\bar{t}/(1-\rho_0)$. When the parameters vary, the maximum load will therefore satisfy the equation

$$\frac{L_{\max}\bar{t}}{1-\rho_0}=\frac{\rho_{\max}-\rho_0}{1-\rho_0}=\text{constant} \tag{8.3}$$

This equation gives the maximum load for the value $\rho_{\max}'$ when the initial value is known. It also provides $L_{\max}'$.

In fact, it is difficult to justify this empirical approach by an exact calculation, at least within the framework of interrupt systems. It is best to keep ρ_{max} constant.

Let us then assume an improvement in the software which is reflected by the modification of the t_i in eqn (8.2) and therefore in $\bar{t}$. Then,

$$L_{max}' = \frac{L_{max}\bar{t}}{\bar{t}'} > L_{max}$$

Similarly, a change in the composition of the traffic (e.g. an improvement in the efficiency of the network) is reflected by a modification of the f_i and in this case an increase in $\bar{t}$; L_{max} decreases. The change in the dialing system and the progressive introduction of keypad sets similarly modify the f_i.

In the case of expansion of the exchange, ρ_0 will vary:

$$\rho_0' = \rho_0 + \alpha(N' - N) + \beta(J' - J)$$

and L_{max} will decrease.

These changes can be represented graphically (Fig. 8.3).

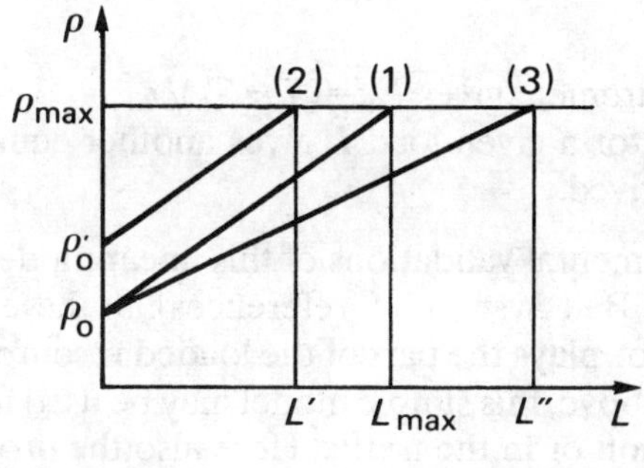

Fig. 8.3 Effect of the modifications on the capacity: line (1), initial model; line (2), effect due to change in ρ; line (3), effect due to change in $\bar{t}$.

Note that there are experimental verifications of the linear model (see references).

8.4.1 Cyclic type of control

The model developed above implies in particular to a processor controlled by interrupt levels. When the control of telephone tasks is entrusted to a cyclic program, a similar analysis may be carried out; it is then the cycle time which is of key importance.

The following analysis therefore assumes a cyclic type or mixed type of processing, i.e. tasks with time constraints attached to an interrupt level (clock) and call processing according to a defined cycle (cf. previous chapter). We then write for T, which denotes an observation period (e.g. busy hour),

$$T = T_0 + L\bar{t} + \tau E \tag{8.4}$$

where T_0 is the constant time spent by the scanning programs (corresponding to ρ_0), $\bar{t}$ and L have the same meanings as previously, τ is the time spent by the supervision (the overhead) of the cyclic program and E is the number of cycles carried out in T.

$T - T_0$ is the time available for call processing. The grade of service is related to E. T/E is the time that it takes for 1 cycle. Preselection requires a large number of tasks to be accomplished and therefore a large number of cycles and thus a value $E_{\min}$ exists which makes the preselection time prohibitive.

We then have

$$L_{\max} = \frac{T - T_0 - \tau E_{\min}}{\bar{t}}$$

The measurement of T_0, τ and $\bar{t}$ gives rise to the same problems as they do in interrupt control. However, here we are measuring E, the number of cycles per unit time.

- The no-load measurement gives $E_0 = (T - T_0)/\tau$.
- The measurement for a given load L gives another equation from which the parameters can be derived.

There are experimental validations of this linear model using the ESS1 and TSPS1 systems of the Bell System (cf. references); in these centralized systems, of course, the processor plays the part of the loaded resource.

Similarly to the above, this simple model may be used to predict the effect of changes in the operation or in the traffic. Here also the problem of the relationship between $E_{\min}$ or $L_{\max}$ and the exchange parameters arises. The same approach as that which produced eqn (8.3) makes it possible to postulate that

$$\frac{T - T_0}{\bar{t}L_{\max}} = \text{constant}$$

This is a relationship which is just as questionable as eqn (8.3) and which provides $E_{\min}/\bar{t}L_{\max} = \text{constant}$.

8.4.2 Conclusion

We have presented a very simple linear model here. Of course, this model is limited. In particular, the above-mentioned measurements justify such a model for processors with cyclic or interrupt control over a broad range of variation. However, a validation for some centralized systems, some of which are rather old, cannot guarantee its universal application! How is it possible to choose ρ_{max} or E_{min}? We have related these parameters to the preselection time. In fact, in general there are several constraints which the exchange has to satisfy. Each parameter (the preselection time and selection time for each type of call) gives rise to a limitation of the utilization rate of each resource. Following [Farber 70], we define the main constraint of service as follows: if we denote $i = 1, 2, \ldots, I$ as the resources constraints, $j = 1, 2, \ldots, J$ the service grade constraints, we have

$$\max_i \max_j (\rho_{i,j}) \leqslant \rho_{max}$$

Finally, the most critical point remains the determination of ρ_{max} or E_{min}. We have discussed the relationship between ρ or E and the preselection time but it is difficult to relate the preselection specification to a limiting load on the processor. In order to do this, it is necessary to construct a more detailed model of the control system and to invoke the mathematical apparatus and the concepts of queuing theory.

The solution of this detailed model (when it is possible) will define the limits within which the linear assumption can be applied.

It will be seen in Chapter 10 that the linear model can be used as a qualitative model for interpreting overload phenomena.

References

In [Villar 77] a well-documented survey of the problems and methods of traffic evaluation is presented; it contains numerous additional references.

As far as experimental proofs of Poisson arrivals and stochastic assumptions are concerned, relatively few sets of measurements are reported in the literature because of their cost. [Dietrich 75], in which a traffic model is also developed, and [Myskja 73] should be consulted. In [Anderberg 76] and [Roberts 79] the data in Section 8.2 is completed. For data concerning abnormal situations (overload), see Chapter 10.

Possible optimizations of the SPC exchange capabilities are discussed in [Potter 73].

The first reference to "environment simulation" applied to traffic estimation is [Fontaine 71]. Case studies are described in [Gruszecki 76] (for the METACONTA 10C), in [Man 73] (for ESS1 overload control), in [Lefebvre 81] (for E12) and [Fiche 81] (for E10B).

The linear model was first proposed in [Farber 70]. Experimental verifications have appeared in [Jaegger 72] (for ESS1 and TSPS1 and eqn (8.3) is taken from this reference), [Tanno 74] and [Lefebvre 81]. In [Brand 77] the extension of cyclic and "pure timed" schedules is discussed.

[Anderberg 76] M. ANDERBERG, G. WIKELL. Call generation, holding times and traffic load for individual telex and telephone lines. *Proceedings of the 8th International Teletraffic Congress*, Melbourne, 1976, Paper 635.

[Brand 77] J. E. BRAND, J. C. WARNER. Processor call carrying capacity estimation for SPC systems. *Proceedings of the IEEE*, Vol. 65, No. 9, September 1977.

[Dietrich 75] G. DIETRICH. Traffic model for common control switching systems. *Electrical Communication*, Vol. 50, No. 1, 1975, pp. 28–34.

[Farber 70] N. FARBER. A model for estimating the real time capacity of certain classes of central processors. *Proceedings of the 6th International Teletraffic Congress*, New York, 1970, Paper 426.

[Fiche 81] G. FICHE. Methods for performance evaluation of the E10B system (in French). *Commutation et Transmission*, No. 4, 1981, pp. 69–76.

[Fontaine 71] B. FONTAINE. Real time environment simulation. *Electrical Communication*, Vol. 46, No. 3, 1971, pp. 188–190.

[Gruszecki 76] M. GRUSZECKI. ENTRASIM, a real traffic environment simulator for SPC switching systems. *Proceedings of the 8th International Teletraffic Congress*, Melbourne, 1976, Paper 435. Also *Electrical Communication*, Vol. 53, No. 1, 1978.

[Jaegger 72] R. J. JAEGGER, R. L. POTTER. Analysis of processor usage in SPC telephone switching systems. *Proceedings of the International Conference on Communication*, Philadelphia, 1972, Paper 16.

[Lefebvre 81] B. LEFEBVRE, J. LAMY, J. P. CREMIEUX. E12 digital transit centers: testing procedures and measured performance data. *Proceedings of the International Switching Symposium*, Montreal, 1981, Paper 24A2.

[Man 73] F. T. MAN. Real time testing of automatic overload control system in a laboratory environment. *IEEE Transactions on Communications*, Vol. COM 2, No. 9, September 1973, pp. 1027–1031.

[Myskja 73] A. MYSKJA, O. O. WALMANN. An investigation of telephone user habits by means of computer techniques. *IEEE Transactions on Communications*, Vol. COM 21, No. 6, 1973, pp. 663–671.

[Potter 73] R. L. POTTER. Getting the most out of stored program control. *Bell Laboratories Record*, June 1973, pp. 181–185.

[Roberts 79] J. ROBERTS. Recent observations of subscribers' behaviour. *Proceedings of the 9th International Teletraffic Congress*, Torremolinos, 1979.

[Tanno 74] T. TANNO, S. TAKAMURA. K. ISHII. Operating experience gained from the D10ESS. *Proceedings of the International Switching Symposium*, Munich, 1974.

[Villar 77] J. E. VILLAR DE VILLACIAN. Traffic calculation in SPC systems. *Electrical Communication*, Vol. 52, No. 3, 1977, pp. 199–205.

Chapter 9

ANALYTICAL MODELS OF THE CONTROL SYSTEM

The application of queuing theory to performance evaluation of the control system is covered in this chapter. This chapter and the previous chapter should be read together as they are part of the same topic.

It should be noted that there is an important difference from the subject covered in Chapter 4 as the existing architectures are too diverse for there to be any systematic approach. We shall study separately the elements of the mechanism already discussed in Chapter 7; the control system consists of a combination of these elements.

We should add that mathematical difficulties will limit the scope of our arguments and make it necessary to resort to approximations.

We start by constructing a queuing model of the control system in Section 9.1; the performances (preselection and selection delays) can be derived from this, most often by isolation and separate treatment of the components.

Section 9.2 shows the type of results which the theory gives for simple servers (peripheral units); these are the mean and variance of the waiting times.

Processors are covered in Section 9.3. They can be organized in a large number of ways; we look at the main ones.

Finally, when the problems are insurmountable or in order to test the inevitable assumptions, simulation will be relied upon considerably. We shall review in Section 9.4 some aspects of its application within this framework.

Throughout this chapter, we shall be forced to maintain a reasonable theoretical level while looking for models sufficiently universal to be usable and results sufficiently powerful to be of some help. This consideration has led to the discarding of results which are too theoretical (the references complete this topic). The methods presented here represent all those which can be used in practice by the teletraffic engineer at this time. Progress remains to be made (and is being made!) in this area.

Let b, s and w denote the durations of service, of sojourn and of waiting at a station. These are random variables for which the first moments are $E(b)$, $E(s)$ and $E(w)$ with the corresponding notation b, s and w, the second moments are

$E(b^2)$, $E(s^2)$ and $E(w^2)$, and $\sigma_b^2 = E(b^2) - b^2$, $\sigma_s^2 = E(s^2) - s^2$ and $\sigma_w^2 = E(w^2) - w^2$ are the standard deviations.

The coefficient of variation of a distribution is the ratio c of the standard deviation to the mean, e.g. $c = \sigma_b / b$.

9.1 A global model of the control system

Telephone calls can be regarded as customers receiving a service in the stations and then leaving them for other stations. Chapter 7 has shown that a queuing model of the control system can be constructed in this way; in this the calls evolve according to specific routes.

9.1.1 A centralized control system

In a centralized architecture, the network of queues takes the very simple form of the central server network (Fig. 9.1).

Each circle symbolizes a station (one or more servers operating with load sharing); this is preceded by its queue possibly with limited capacity. The single central server (the control unit) is of a special type. Several queues feed it; the customers which it serves do not have the same priority or the same characteristics.

The number of queues in front of the control unit depends on the organization. In an interrupt control system, each level feeds a specific queue. The service discipline (priority control (preemptive or not), control of the cyclic type (where the server scans the queues in a defined order etc.). i.e. the choice of the queue for selecting the customer to be served, also depends on the operating mode of the scheduler (described in Section 7.4).

We recall that a priority is preemptive if it interrupts the task in progress in order to take the server. It is assumed that, at the end of execution of the

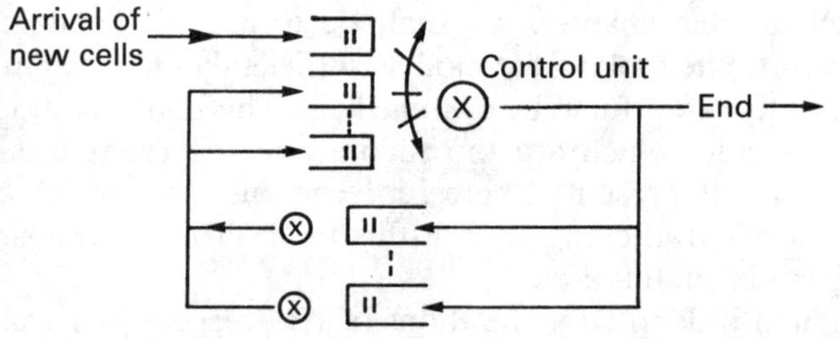

Fig. 9.1 Central server model of the centralized control system.

interrupting task, the interrupted task is resumed at the point where it was stopped (preemptive resumption).

The method of linking the control unit and the periphery takes various forms (cf. Section 7.2). The bus, the physical connection medium between the peripherals and the central unit, is not generally involved. Telephone systems usually have a specialized bus whose load remains small and does not affect the operation; the exchange of information is usually controlled by the processor which manages the bus. Therefore, this cannot give rise to conflict or congestion (contrary to the case in noncentralized systems, e.g. "contention" bus). According to the case, the processor controls the exchange by polling (it gives cyclically each unit the right to transmit and send its orders) or it controls the exchange by interrupts (on a signal from the periphery, it takes the information from there). This operating mode also applies to systems of the decentralized type.

Where a "controller" exists or if it is desired to take into account the transfer time and the waiting time for the bus, an intermediate station will be added as a buffer between the control unit and the peripherals.

The routing of customers in the network is described by the flow diagram for the routing of the calls (see Section 7.3). This is a fixed path for each call mix category (the specialists refer to "deterministic routing") and will vary from one category to another.

9.1.2 A distributed architecture

The principle of modeling by a network of queues can still be applied as well as the remarks of the previous section. However, the network obtained is no longer a central server network and does not take a "universal" form. A system of the type "universal core" will give rise to multiserver stations (control processors operating with load sharing) for which it is necessary to define the rules for accessing the common incoming queues. A system with "functional distributed core" will have several central servers in cascade etc.

The most general distributed system will give rise to a general queuing model. Most frequently, its peripherals will be complex (each peripheral in this way can be made a central server model; this is already the case for control systems of the decentralized type, e.g. the regional processors of the AXE system). In practice, the model must be constructed by the superimposition of the submodels of each component.

Note that a control system of the decentralized type will give rise to a hybrid of the two types of network, i.e. a central server network in which each peripheral station is a central server model.

In general, the complete model must be simplified. In order to do this, a procedure of "breaking down and building up" is adopted. A peripheral is isolated whose separate model gives the characteristics (output flow and transit time); it is thus possible to build up by constructing a model of the peripheral equivalent to a single queue which will then be introduced into the global model.

9.1.3 Calculation of the flows

In summary, it is always possible to construct a queuing model which is realistic to a large extent, in particular, if allowance is made for the time schedule of the processors. The model receives a stream of calls from the outside (the arrivals of requests based on the traffic model in Chapter 8) which cause known occupancies in the stations, since the routing of each incoming customer is known.

It can therefore be used directly for predicting the capacity and the performances of the system by simulation.

However, some problems will arise for the direct and exact application of the calculation methods. In practice, approximate methods will have to be adopted.

The model in Fig. 9.1 will provide the flows passing through each station. A conservation equation will be applied which can be immediately interpreted.

We call Λ_i, $i=1, \ldots, I$, the flow of outside arrivals for each type of communication (the number of incoming calls per unit time). Each type i follows a complex but known path leading it $k_{i,n}$ times to the station n ($n=1, 2, \ldots, N$).

The flow of customers passing through the station n will simply be

$$\lambda_n = \sum_i \Lambda_i k_{i,n} \tag{9.1}$$

This is the total number of customers received per unit time at n.

Another description can be used in which the routing of the customers will be random. The client who leaves station m is directed to station n with probability $q_{m,n}$; usually a $(N+1)$th imaginary station is added which functions as the output. $q_{n,N+1}$ is the probability of leaving the network; similarly, $q_{0,j}$ is the probability that an outside arrival is directed to j.

The call-processing flow diagrams establish the parameters of this representation; for the type i of call, let $h(i, n, m)$ be the number of transitions from station n to station m. Then,

$$k_{i,m} = \sum_{n=0}^{N} h(i, n, m) = \sum_{n=1}^{N+1} h(i, m, n)$$

$$q_{m,n} = \frac{\sum_i \Lambda_i h(i, m, n)}{\sum_i \Lambda_i k(i, m)}$$

In following the usual concepts and terminology of queuing network theory (see Section 9.1.4 below), the concept of customer class is introduced. The routing of a customer, i.e. the q, becomes a function of its class; the changes in class are made at the output of the stations. The following coefficients are defined: $q_{j,r;k,s}$ is the probability that a customer of class r who leaves station j is directed to the station k with class s; $q_{0;j,r}$ is the probability of entering the network by the station j with class r; $\lambda_{j,r}$ is the flow of class r passing through the station j, λ_j the total flow to the station and Λ the total flow in the network.

$$\lambda_{j,r} = \Lambda\, q_{0;j,r} + \sum_{k,s} \lambda_{k,s} q_{k,s;j,r} \tag{9.22}$$

$$\lambda_j = \sum_r \lambda_{j,r}$$

The concept of class unifies the preceding descriptions and generalizes the description of the type of call; for a deterministic path, a set of classes is defined for each type of call which guides the customer (along the path, q is only made equal to 0 or 1).

The network of queues will have a stable behavior if the utilization rate of each server remains less than unity. $b_{m,s}$ denotes the average service time, which is possibly a function of the class (i.e. type of call), and R_m is the number of servers of station m:

$$\rho_{\mathrm{m}} = \frac{\sum_s \lambda_{m,s} b_{m,s}}{R_m} < 1 \text{ for } m = 1, 2, \ldots, N \tag{9.3}$$

This condition can be directly verified from the data of the architecture and call processing, the operational data and the data of the traffic model by assuming that the waiting phenomena do not modify the parameters; this is not the case under overload conditions in particular where the overload control acts by reducing the $b_{i,r}$ and where the λ vary.

If we impose a limit on the utilization rate (e.g. $\rho_i < 0.7$ for all i), eqn (9.3) allows us to calculate the permissible capacity; in fact, we return to the deterministic model of the previous chapter.

The results expected here from the study of the control system are of two types.

- Offered grade of service, i.e. the performances of the machine quantified by the preselection and selection times and the through-connection delay and traffic capacity, and the value of the load making the service grade prohibitive (more precisely than the simple requirement $\rho < 0.7$ in Chapter 8).
- An aid in the dimensioning of the memories, buffers and peripheral units.

The evaluation must be able to measure the distributions of the number of customers and especially the response time of a customer (the time elapsing between its arrival and exit from the model). It would be worth being able to calculate the distribution of these times or at least the first moments.

9.1.4 Application of the Baskett–Chandy–Muntz–Pallacios theorem

The exact analytical methods which can be currently used on queuing networks reduce to the Baskett–Chandy–Muntz–Pallacios (BCMP) theorem. This theorem makes it possible to calculate the state probabilities under steady-state conditions (the probabilities of finding $n_1, \ldots, n_N$ customers in the queues $1, 2, \ldots, N$), making some rather strong assumptions.

- Servers of the exponential type and service in the order of arrivals or any service distribution but a discipline of the type "last in, first served" with preemption or processor sharing or a station with an infinite number of servers (here also any distribution). In the "processor-sharing" discipline the server shares its time equally between the customers who are present; if there are n of them, each receives $1/n$ s per second. This is a good idealization of the time-sharing schedule policies.
- Routing governed by a Markov chain, possibly as a function of the customer class. This assumption enables deterministic routing to be taken into account without any error as was noted above (the customers change class each time that they pass through the central server); however, there is an increase in the number of classes.

By means of these assumptions, the theorem provides the probability distribution for the number of customers in the network. If we denote $n_1, n_2, \ldots, n_N$ as the number of customers at stations $1, 2, \ldots, N$, the distribution is of a "product form", i.e. it has the form

$$P(n_1, n_2, \ldots, n_N) = P_1(n_1)P_2(n_2) \ldots P_N(n_N)$$

The BCMP theorem therefore states that, as far as the state probabilities are concerned, everything happens as if the stations were independent, the probability of observing the state $n_1, n_2, \ldots, n_N$ being exactly the product of the marginal probabilities $P_i(n_i)$ which are calculated as if the server i were isolated and supplied by the Poisson flow with the intensity given in eqn (9.2).

The dimensioning of the queues is derived from these probabilities. Unfortunately, despite the restrictions that it imposes, the theorem does not completely answer the questions; it does not yield the sojourn times in the network. The "product form", which is exact for occupation probabilities, does not extend to the sojourn time; the stations are not independent. At the most, it allows the average sojourn time at each station to be evaluated separately using the Little formula.

The references should be consulted for a complete statement of the theorem and its proof [Lavenberg 83].

9.1.5 An isolation method

The rather restrictive assumptions of the BCMP theorem make it necessary to find other solutions, which are necessarily approximate. The most immediate method consists in assuming the following.

- Independent stations. The sum of the means and variances of sojourn times then provides the mean and the variance of the total time.
- A Poisson character for all the flows which supply them, the intensity of these flows being given by eqn (9.1) or (9.2).

Each station is treated as an $M/G/1$ queue, possibly with priorities, abandoning the restrictive assumptions of the BCMP theorem. The distribution of the total time will be the convolution of the time distributions of each station as if they were independent. Usually, the first moments suffice; the mean time and the variance of the total time will be the sum of the means and the variances.

Experience (comparison with simulation) shows that this approximation gives excellent results, in particular if the network satisfies the assumptions of the BCMP theorem.

The approximation is therefore made as follows. Let $i = 1, 2, \ldots, n$ be the stations traversed by the call of the selected type (e.g. successful outgoing preselection), t_i and σ_i be the mean and the standard deviation of the sojourn time at station i and k_i be the number of passages to the station i. T and σ are the mean and standard deviation of the time selected. Then,

$$T = \sum_i k_i t_i$$
$$\sigma^2 = \sum_i k_i \sigma_i^2 \tag{9.4}$$

because of the assumed independence. For successive passages through the control processors, these will generally be assigned different priorities and the parameters t_i and σ_i of successive passes will differ.

Equations (9.4) may even provide the quantiles of the required time by using the Martin approximation (cf. Appendix at the end of the book). These quantiles are fundamental in fixing the nominal capacity.

It remains to calculate the t_i and σ_i of the different servers. An "isolation" type of approach is used. The performances are evaluated by isolating each station of the network and by supplying it with a Poisson process. This then amounts to a study of a simple queue with one or more servers, which can be more easily solved (simulation is a possible method). Section 9.2 will deal with peripherals and Section 9.3 with processors.

9.1.6 Other global approximations

The decomposition approximation of the previous section can be strengthened by characterizing the arrival process in terms of two parameters. Kuehn suggests such an approach (see [Kuehn 76] and [Kuehn 79]). He adopts the "isolation principle" already used above; each component (which may be itself a subnetwork) is studied independently of its neighbors.

In contrast with the analysis presented above where the flows offered to the stations were Poisson, Kuehn represents them in terms of two parameters (the mean and the variance of the time between arrivals).

He then assumes that each station is of the type $GI/GI/1$ which he handles by means of an approximation which makes it possible to relate the parameters of the incoming and outgoing flows. He also obtains the average sojourn time. The approximation retains only the first two moments of the service distribution.

It is also necessary to solve the problem of combining the flows, a problem which is not solved satisfactorily in the opinion of Kuehn himself.

The analysis therefore provides the average sojourn time of a customer in the network, taking into account the flow characteristics better than in Section 9.1.5. Nevertheless, this approach ignores the concept of the type of customer (it averages the characteristics of the different types of customer) and does not give any information on the variance of the sojourn time.

Reference should also be made to the discussion of Labetoulle and Pujolle on isolation methods in [Labetoulle 80]. All these methods are of the iterative type. The equations for the stations relate the flow parameters (the mean and the variance) to each other; in the general case, a "closed" algebraic solution is not obtained with the exception of the means (eqn (9.2)).

Another approach to this problem consists in using the technique of approximations by diffusion processes where the discrete process (arrival and departure of the customers) is replaced by a continuous process. Here also, the distributions are taken into account in terms of their first moments.

All these more sophisticated methods turn out to be proportionately more difficult to explain and to use. They will not be developed here since [Lavenberg 83] and [Gelenbe 80] provide an excellent presentation. They can be applied directly to the control system of exchanges.

Finally we point out that most queuing analysis packages take them into account.

9.2 The peripherals

The peripherals are simple units directly amenable to the queuing model in Fig. 9.1 and the calculations presented below, or more complex units organized around a local processor; in the latter case the entire peripheral is a central server system to which the approach is applied.

In the following, they will be studied separately from the central control system by assuming that they are supplied with a Poisson stream with intensity given by eqn (9.2).

The customers are usually served in the order of their arrival. The service time may vary according to the type of call and the state of the peripheral. A constant time will nonetheless be a good approximation in general even if certain studies make intensive use of the exponential distribution.

We shall give an example of the type of result obtained by queuing theory. It certainly does not cover all the possible configurations.

The communication mode between the peripheral and the processor must be taken into account.

9.2.1 Interrupt link

When the unit has terminated its work, it interrupts the processor which initializes the reading of the result and the sending of the next order. The time spent in controlling the transfers is integrated into the total processing time (Fig. 9.2).

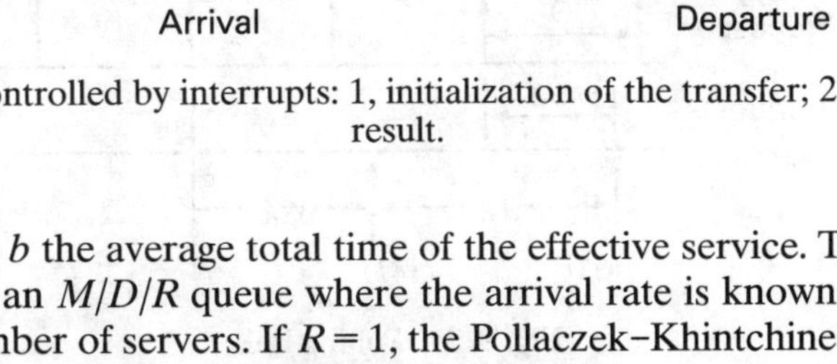

Fig. 9.2 Link controlled by interrupts: 1, initialization of the transfer; 2, reading of the result.

Let us call b the average total time of the effective service. The peripheral is equivalent to an $M/D/R$ queue where the arrival rate is known (eqn (9.2)). R denotes the number of servers. If $R = 1$, the Pollaczek–Khintchine formulas give the mean and the variance of the waiting time:

$$W = b\frac{\rho}{2(1-\rho)} \tag{9.5}$$

$$\sigma_w^2 = \frac{\rho b^2(4-\rho)}{12(1-\rho)^2}$$

where $\rho = \lambda b$. The distribution is also known:

$$W(t) = P(\text{waiting time} \leqslant t)$$

$$= (1-\rho)\, \mathrm{e}^{\lambda t} \sum_{i=0}^{k} \frac{\mathrm{e}^{-ib\lambda}[(ib-t)\lambda]^i}{i!} \qquad \text{for } kb \leqslant t \leqslant (k+1)b$$

9.2.2 Polling link

The scanning is periodic (polling), leading to a waste of time for the peripheral whose effective service time is an integral number of cycles.

We call θ the cycle time, h the time of operation of the unit and b the effective service time (Fig. 9.3):

$$b = \theta \left[\frac{h}{\theta} \right]$$

where $[x]$ designates the integer equal to x or immediately greater than x. We call r_k the time which elapses between the arrival of the request k and the next cycle and t_k the time separating the arrivals of the requests $k-1$ and k.

The request k undergoes a waiting time given by $w_k = r_k$ if the peripheral is free, and by $w_k = w_{k-1} + b - t_k$ otherwise.

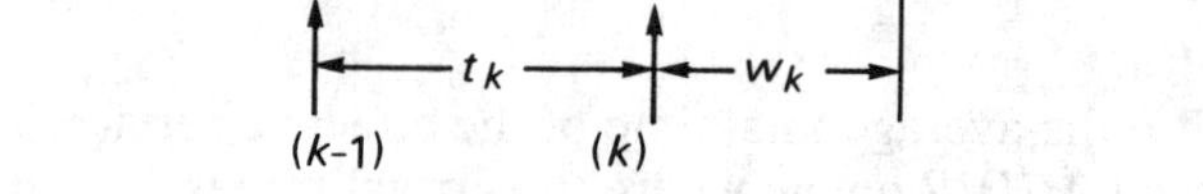

Fig. 9.3 Polling link.

This is therefore an $M/D/1$ queue where the customer finding an idle server has a particular service time. An analysis of this has been made using Laplace transforms. The stability condition is always $\rho = \lambda b < 1$. We call η the probability that an arrival does not have to wait but undergoes the "dead time" of the server. We obtain

$$P(r_k \leqslant t) = \frac{e^{\lambda t} - 1}{e^{\lambda \theta} - 1} \qquad \text{if} \quad 0 \leqslant t \leqslant \theta$$

$$\eta = \frac{(1-\rho)(1-e^{-\lambda\theta})}{\lambda\theta}$$

The average time is

$$E(r) = \frac{\theta}{1-e^{-\lambda\theta}} - \frac{1}{\lambda}$$

The Laplace transform of the waiting time is

$$f(s) = \int_0^\infty e^{-st}\, dW(t) = \frac{1-\rho}{\theta}\, \frac{1-e^{-s\theta}}{s-\lambda+\lambda e^{-sb}}$$

from which we derive the mean time and the variance

$$W = -\frac{d}{ds}\{f(s)\}\Big|_{s=0} = \frac{\theta}{2} + \frac{b\rho}{2(1-\rho)}$$

$$\operatorname{var}(w) = \frac{d^2}{ds^2}\{f(s)\}\Big|_{s=0} - W^2 = \frac{\theta^2}{12} + \frac{\rho b^2(4-\rho)}{12(1-\rho)} \qquad (9.6)$$

The analysis outlined here is based on [Gerrand 76] to which reference should be made for the details of the calculations.

9.3 The processors

They will need a more refined and mathematically more arduous analysis because of their specific characteristics. Several types of task are offered to them with different characteristics (arrival and service processes), serviced according to a complex rule (control by priority or of the cyclic type) (cf. Section 7.4). Care should be taken that it is a question here of interrupts for scheduling the tasks of the processor and no longer for input–output control. Moreover, the two types may cohabit.

- A class of particular interrupts often exists, e.g. the clock cycle with absolute priority.
- The flows which occur are not independent, i.e. a task leaves the server in order to return in another class; this is a feedback queue.

9.3.1 A model of the processor

In the general case, the processor is a unit which serves several queues according to a priority rule (see Section 9.3.2) or a fixed cycle (see Section 9.3.3). The customers which feed it are of two types.

- Outside arrivals (new calls) which form a Poisson flow. Very often, the device for admitting the calls linked to the clock interrupt transforms this flow into group arrivals. If we denote T as the clock period and λ the traffic, the size of the group is given by

$$P(\text{size} = k) = \frac{(\lambda T)^k}{k!} \mathrm{e}^{-\lambda T} \qquad (9.7)$$

- Internal feedbacks, i.e. processed tasks switched to a peripheral and returning to the control unit. In Section 9.1, we represented the resulting flow by Poisson traffic; this will be incorrect especially for the central server.

In the ideal case where the assumptions of the BCMP theorem apply, feedback does not pose any particular problem; everything happens for the occupation probabilities as if the server were an $M/M/1$ queue. The waiting times can be derived from this by the Little formula.

Another interesting case is that of "instantaneous" feedback, i.e. a customer leaving the server goes back to the queue. Thus the multiprograming applications require an algorithm in which each task present receives a quantum of processor time before returning to the queue. The processor is shared equitably between all the programs present. The "processor-sharing" discipline is the idealization of this type of scheduling (see Section 9.3.5). The group arrival model can also be used (see below).

More elaborate analyses of instantaneous feedback have been made; they take into account the feedback and the existence of priorities (see for example [Enns 69], [Enns 70], [Chan 73], [Haugen 76] and [Altamirano 83]). These analyses, which are very complex in terms of the mathematics, will not be presented here.

In any event, instantaneous feedback seems rather a rare solution and it is necessary to consider feedback with delay.

All the models suggested in the literature on this subject assume a processor with priorities and instantaneous feedback. Several practical solutions are available to the modeler for the study of systems not satisfying these assumptions. None is really satisfactory.

- Neglect of the feedback phenomenon by assuming a Poisson stream whose intensity is given by eqn (9.2). According to the discipline for controlling the processor, Sections 9.3.2 and 9.3.3 deal with the problem.

- Neglect of the existence of the delay before feedback and application of existing models. These models are mathematically arduous and without doubt rather specific.
- Modification of the task arrival model. Since the external arrivals take place (in most systems) by groups at the beginning of each clock cycle, but since the feedbacks are caused in the final analysis by these arrivals, the composite stream offered is replaced by a single stream defined as follows Fig. 9.4.
  - The arrivals occur on clock pulses;
  - The arrivals occur in each queue by groups;
  - The size of the groups is such that it preserves the load of each queue (it is given by eqn (9.7) for queue 1; for the others, the average size is $\lambda_i T$).

In Section 9.3.4 we shall return to this model in connection with the clock interrupt.

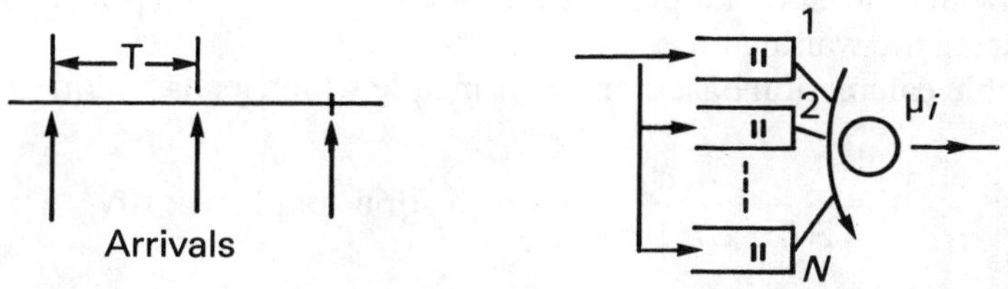

Fig. 9.4 Equivalent model of the arrivals at the processor.

9.3.2 Priority control

Most real-time systems make significant use of priorities with or without interrupts. Cyclic interrupts will be examined separately.

This control is modeled as an $M/G/1$ queue with priorities. We speak of "head of the line" priority. The changeover time (taking into account the interrupt of software control of the priorities) is included in the service.

Without giving a complete theory, we shall succinctly show the principle of the solution. The priority levels are numbered 1 to P, the class j having priority over $j+1, j+2, \ldots, P$.

The sojourn time for a customer is separated into the waiting time and the service time. The latter is formed from the service time proper and, for preemptive priorities, the service time of priority clients interrupting it. We then denote $\tilde{b}_i$ as the average effective service time including interrupts and

$$\tilde{b}_i = b_i + \tilde{b}_i \sum_{k=1}^{i-1} \lambda_k b_k$$

Let

$$\tilde{b} = \frac{b_i}{1 - U_{i-1}} \tag{9.8}$$

where

$$U_k = \sum_{j \leqslant k} \lambda_j k_j$$

U_{i-1} denotes the load on the processor produced by the levels of priority higher than i. Everything happens as if the service time were prolonged by the unavailability of the server. If the priority is simple, $\tilde{b}_i = b_i$.

The waiting time is separated in a similar manner into the remaining service time of the customer occupying the server if the priority is nonpreemptive or if the customer has at least the same priority, the service time of waiting customers with at least equal priority and the service time of priority customers arriving during the waiting time.

A simple calculation based on the principle which leads to eqn (9.8) gives

$$W_i = \frac{S(P)}{2(1 - U_i)(1 - U_{i-1})} \qquad \text{for a simple priority} \tag{9.9}$$

and

$$W_i = \frac{S(i)}{2(1 - U_i)(1 - U_{i-1})} \qquad \text{for a preemptive priority} \tag{9.10}$$

where we have set

$$S(k) = \sum_{j \leqslant k} \lambda_j E(b_j^2)$$

The total sojourn time is $s_i = W_i + \tilde{b}_i$.

A modification of the calculation makes it possible to deal with the very frequent case of a mix of preemptive and nonpreemptive priorities. Each level i is broken down into sublevels $j = 1, 2, \ldots, m$ so that a task (i, j) interrupts the tasks (i', j') if $i' > i$ but has a simple priority over a task (i, j'), $j' > j$. The number of sublevels is assumed to be the same for all i without any restriction on generality (imaginary sublevels can be added).

The total residence time is always the sum of the waiting time and the effective service which is separately evaluated:

$$\begin{aligned} s_{i,j} &= W_{i,j} + \tilde{b}_{i,j} \\ \sigma_s^2 &= \sigma_w^2 + \sigma_b^2 \end{aligned} \tag{9.11}$$

The notation is simplified by symbolizing the state (i, j) by ν. Thus, U_ν replaces $U_{i,j}$ and $U_{\nu-1} = U_{i,j-1}$. Finally, we continue to use U_i and $S(i)$ and we introduce $V(i)$ for quantities related to complete levels. Thus,

$$U_i = \sum_{k=1}^{i} \sum_{n=1}^{m} \lambda_{k,n} b_{k,n}$$

$$S(i) = \sum_{k \leq i} \sum_{n \leq m} \lambda_{k,n} E(b_{k,n}^2)$$

$$U_\nu = U_{i,j} = U_{i-1} + \sum_{k=1}^{j} \lambda_{i,k} b_{i,k}$$

$$S(\nu) = S(i-1) + \sum_{k=1}^{j} \lambda_{i,k} E(b_{i,k}^2)$$

$$V(i) = \sum_{k \leq i} \sum_{n \leq m} \lambda_{k,n} E(b_{k,n}^3)$$

Then, for a customer of class $(i, j) = \nu$, we have

$$W_{i,j} = W_\nu = \frac{S(i)}{2(1-U_\nu)(1-U_{\nu-1})}$$
$$E(W_{i,j}^2) = \frac{V(i)}{3(1-U_\nu)(1-U_{\nu-1})^2} + \frac{S(i)S(\nu)(2-U_\nu-U_{\nu-1})}{2(1-U_\nu)^2(1-U_{\nu-1})^3} \tag{9.12}$$

The effective service times become

$$\tilde{b}_\nu = \frac{b_\nu}{1-U_i}$$
$$E(\tilde{b}_\nu^2) = \frac{E(b_\nu^2)}{(1-U_{i-1})^2} + \frac{b_\nu S(i-1)}{(1-U_{i-1})^3} \tag{9.13}$$

It will be even possible to extend these formulas to any organization of priorities and "preemption distances"; the results then look like numerical algorithms (see [Herzog 75] and [Herzog 76]).

For the proofs of the formulas presented here, reference should be made to the classical studies in [Jaiswal 68], [Kleinrock 75] and [Gay 75].

9.3.3 Control of the cyclic type

The processor serves I queues according to a fixed cycle; Fig. 7.4 illustrates the cycle of the central processor of ESS1. It has a length of 30 for 5 queues to be served.

Depending on the case, the processor will entirely empty a queue before passing to the next one (exhaustive service) or will only serve a limited number on each pass.

The changeover from one queue to the next takes a non-zero processor time but shorter than the handling of an interrupt; it is assumed to be non-zero for practical reasons (see eqn (9.14)).

The analysis of this operation turns out to be extremely difficult and no final solution has yet been found (note the analogy to the multiplexing of transmission lines).

$i=1, 2, \ldots, I$ denotes the queues to be served, λ_i denotes the flow of arrivals at the queue i and b_i the service time ($\rho_i=\lambda_i b_i$); c_i is the average time of passing from the queue i to the next one ($c_i^{(2)}$ is the second moment; it is reasonable to assume that $c_i^2 = c_i^{(2)}$); l_i is the maximum number served on each passage to the queue i (non-exhaustive service). Let q_j be the index of the queue served on the jth stage where $q_j \in \{1, 2, \ldots, I\}$, $j=1, 2, \ldots, N$ (N is the length of the cycle; for the "simple" cycle, $q_j=j$), and n_i is the number of passages per cycle to the queue i ($\Sigma n_i = N$). c_0 is the total changeover time in a cycle (the time of a no-load cycle):

$$c_0 = \sum_{j=1}^{N} c_{q_j}$$

The cycle time is a very important variable. It serves for example to detect an overload. Its mean value is denoted by C. Its equation is easily obtained by invoking the conservation of the number of customers under "permanent conditions"; no queue becomes congested, and all the customers arriving must be served. Each cycle must, on average, carry the number of arriving clients:

$$C = c_0 + (\Sigma \lambda_k C b_k)$$
$$C = \frac{c_0}{1 - U_I} \tag{9.14}$$

where U has the same meaning as in the previous section. It should be noted that a zero changeover time produces a zero average cycle. Under normal conditions, the queues are sometimes all empty; if then the times c_i are all zero, the server accomplishes an infinite number of cycles instantaneously and hence a zero average time.

Exhaustive service

The server empties each queue before switching to the next queue. The stability condition is simply written

$$1 - \Sigma \rho_k > 0$$

It does not depend on c_0 (the changeover time has less and less importance as saturation approaches).

The analysis in the general case provides the Laplace transforms in a form which is difficult to use; [Eisenberg 72] should be consulted. If the cycle is simple ($q_j = j$; $N = I$), the average waiting times are simpler to obtain [Hashida 72]; this method requires nonetheless the solution of a system of I^3 equations. Here we shall use only an approximation of good quality [Bux 83].

The service at queue i can be regarded by a server who is absent for a duration A_i after having emptied the queue ("autonomous server"). The following relationship is exact [Eisenberg 72]:

$$W_i = \frac{E(A_i)}{2} + \frac{\text{var}(A_i)}{2E(A_i)} + \frac{2b_i^{(2)}}{2(1-\rho_i)} \tag{9.15}$$

Let T_i be the time which the server spends in the queue i; $T_i + A_i$ is the duration of the cycle. T_i is the time which the server uses to empty the queue, i.e. to serve the arrivals of the cycle. Thus the mean is given by

$$E(T_i) = \lambda C b_i$$

and hence

$$E(A_i) = (1-\rho_i)C$$

The above formulas are exact. The calculation of the variance turns out to be more complicated so that we make the approximation of independence which allows the variance of A_i to be written as the sum of the variances of T_i (we assume for simplification that the c_i are constant). We then have

$$\text{var}(A_i) = \frac{C}{1-\rho} \sum_{k \neq i} \frac{\lambda_k b_k^{(2)}(1-\rho_i)^2 + \lambda_i \rho_k b_i^{(2)}}{1-\rho_i-\rho_k+2\rho_i\rho_k}$$

It should be noted that this type of control is very vulnerable since the cycle increases indefinitely once one of the queues is congested; this explains the interest in nonexhaustive service.

Nonexhaustive service

The server only serves a limited number on each visit. The saturation of one queue only has a small effect on the neighboring queues, the server not being blocked there.

It is necessary to refine the concept of stability; we shall define a stability condition for the queue k. If this queue is saturated, the server empties l_k customers and eqn (9.14) becomes

$$C = c_0 + C \sum_{j \neq k} \rho_j + n_k b_k l_k$$

The rate of departure from the queue is l_k/C and the maximum arrival rate is simply

$$\lambda_k < \lambda_k^* = \frac{l_k}{C} = \frac{l_k(1 - U_I + \rho_k)}{c_0 + n_k b_k l_k} \tag{9.16}$$

The queues are all stable if $\lambda_i < \lambda_i^*$ for $i = 1, 2, \ldots, I$.

It should be noted that eqn (9.14) only gives the cycle in the case of stability. If all the queues are saturated,

$$C = c_0 + \sum_{i=1}^{N} l_{q_i} b_{q_i} = c_0 + \sum_{i=1}^{I} n_i b_i l_i$$

From this viewpoint, it is equivalent to passing once through the queue i taking l_i customers there or $n_i = l_i$ times taking only one customer.

For this system, as in the previous case, it is difficult to find an exact analysis. We shall limit ourselves to a simple cycle ($q_i = i$ and $I = N$ with $l_i = 1$) and an approximate treatment, which is only valid if $\Sigma\rho_i < 1$ (a more harsh condition than eqn (9.16)).

Under these conditions, [Kuehn 79] provides an approximation (which assumes in fact a certain independence between the queues but is well verified by simulation).

The following auxiliary variables are defined:

$$\sigma_0^2 = \Sigma c_i^{(2)} + c_0^2$$

$$\alpha_j = \lambda_j C$$

$$c_j' = \frac{c_0}{1 - U_I + \rho_k} \qquad \alpha_{j,i}' = \lambda_i c_j'$$

$$c_j'' = \frac{c_0 + b_j}{1 - U_I + \rho_k} \qquad \alpha_{j,i}'' = \lambda_i c_j''$$

α_j represents the probability of a nonempty queue when the server arrives. The variables (α', c', ...) refer to the cycles where a customer from the queue i is served as opposed to (α'', c'', ...) for cycles where no customer from queue i is served. It can be verified that

$$C = \alpha_{j,j}' c_j' + \alpha_{j,j}'' + c_{j,j}''$$

The approximation provides the variance of the cycle

$$\mathrm{var}(C) = \sigma_0^2 - C^2 + (1 - \alpha_j)c_j' + \alpha_j c_j'' + \alpha_j \sigma_{b_j}^2$$

$$+ \sum_{i \neq j} \{(1 - \alpha_j)(\alpha_{j,i}' b_i^{(2)} - \alpha_{j,i}'^2 b_j^2) + \alpha_j(\alpha_{j,i}'' b_i^{(2)} - \alpha_{j,i}''^2 b_i^2)\}$$

as well as the average waiting time

$$W_i = \frac{\sigma_0^2 + \sum_{j \neq i} (\alpha_{j,i}' b_i^{(2)} - \alpha_{j,i} b_i^2)}{2c_i'} + \frac{\left\{\sigma_0^2 + \sum_{j \neq i} (\alpha_{j,i}'' b_i^{(2)} - \alpha_{j,i}''^2 b_i^2) + \sigma_b^2\right\} \lambda_i}{2(1 - \lambda_i c_i'')}$$

The variance is also accessible (see the references).

9.3.4 Clock interrupts

Most systems possess a maximum priority interrupt level for tasks associated with the traffic and activated periodically by a clock. It performs jobs with strong time constraints.

Apart from its interrupt effect described in Section 9.3.2 assuming Poisson arrivals, this level sometimes serves for the admission of new calls (for METACONTA, see Section 7.4; for D10, see below). Tsks activated by the clock often consume a time which does not vary or varies only a little (they are well defined and partly not related to the traffic).

It is extremely difficult to cope with this special characteristic; the difficulty is linked to the nature of the periodic arrival and the mixing with arrivals on other levels. In practice, the modeler has the choice, according to the degree of mathematical complexity tolerated, between the following.

- Neglecting the cyclic nature of the interrupt which reduces to the case in Section 9.3.2. This approximation allows for the load on the processor and extends the waiting time; it can be shown in fact that under constant load a periodic arrival of priority traffic minimizes the waiting of the interrupted flow and is optimum from this point of view. The evaluator finds himself in a secure position.
- Accepting the cyclic nature and studying the response time of the flows 2, …, P interrupted cyclically. Such a study has never been undertaken; for this local model, simulation may provide all the information required.
- Allowing for cyclic interrupts by modifying the arrival process. The difficulty of the preceding approach derives from the mixing of flows with contradictory properties; the model already mentioned in Section 9.3.1 consists in replacing the arrivals in continuous time $\lambda_2, \ldots, \lambda_P$ by producing grouped arrivals at the clock interrupt times.

It is necessary to define the mode of arrival of the packets, i.e. the distribution of their size. On average, $\lambda_i T$ customers arrive between two interrupts; a distribution of the same mean will be used. The geometric or Poisson distributions are often used:

$$P(\text{a group of size } k) = \frac{(\lambda T)^k}{k!} e^{-\lambda T} \qquad \text{(Poisson arrivals)}$$

$$= (1 - \lambda T)(\lambda T)^k \qquad \text{(geometric arrivals)}$$

We are now dealing with a model which does not rigorously copy reality but claims to reproduce its behavior. In the actual system a customer (one task out of the call tasks) makes several trips to the processor and must deal with the competition of tasks from other calls on the same or other priority levels. In the group arrival model, all the tasks generated by a call for this processor are produced at the same time and a customer has to cope with his "brothers" on his level and on neighboring levels. These customers all have the same random characteristics so that it is understandable that the model has some claim to plausibility. In any event, as for every modeling of this type, it is necessary to carry out careful tests by comparing with simulation, for example. This is done in the references.

It should be emphasized that this model, which simplifies the situation, is not sufficient to make it obvious. The method of study consists in observing the system at the clock pulses and writing down the state probabilities at these points. We write down generating functions whose complete determination requires application of the theory of analytical functions!

In [Fujiki 76] the system D10 is studied and the sum of a constant time and an exponential time is adopted as the distribution of the duration of the interrupt. If the clock period T is taken as unit time,

$$P(\text{duration} \leqslant t) = 0 \qquad \text{if } t < \frac{1}{\mu_{10}}$$

$$= 1 - \exp\{-\mu_{11}(t - \mu_{10})\} \quad \text{otherwise}$$

The tasks of levels 2, 3, ... arrive in groups whose size obeys a Poisson distribution and each task has a service time with rate μ_i (on level i).

On completion of a rather arduous analysis, Fujiki and Murao give the average waiting time for a customer of level i:

$$W_i = \frac{1}{2} + \frac{1/\theta_0^{(i)} - 1/\mu_{11} + (1/\mu_i)(1 + \lambda_i/2)}{1 - \sum_{j=1}^{i-1} \rho_j}$$

where $\rho_j = \lambda_j/\mu_j$ and $\theta_0^{(i)}$ is defined by

$$\theta_0^{(i)} = \lim_{z \to 1^-} \{\theta(z)\}$$

where $\theta(z)$ is the solution of

$$0 = 1 - z \frac{\mu_{11}}{\mu_{11} - \theta} \exp\left\{ -\theta \left(1 - \sum_{j=2}^{i} \frac{\lambda_j}{\mu_j - \theta} - \frac{1}{\mu_{10}} \right) \right\}$$

The same model (with constant time of the priority task) is used in [Jung 81] in which an iterative solution is suggested.

9.3.5 The "processor-sharing" discipline

In this ideal model the processor shares its time between all the customers present. As has been already noted, this is the transition to the limit of the service per quanta of time-sharing installations. This may also be a simplified model of the operation with instantaneous feedback (Fig. 9.5).

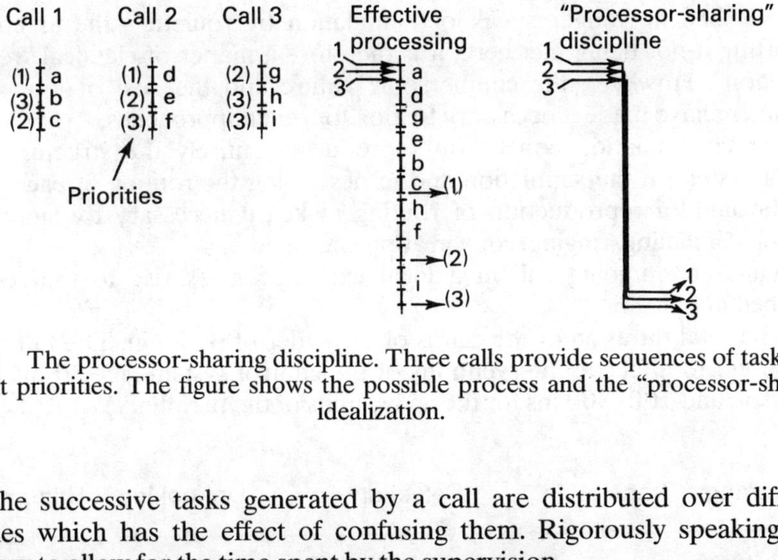

Fig. 9.5 The processor-sharing discipline. Three calls provide sequences of tasks with different priorities. The figure shows the possible process and the "processor-sharing" idealization.

The successive tasks generated by a call are distributed over different priorities which has the effect of confusing them. Rigorously speaking, it is necessary to allow for the time spent by the supervision.

Processor-sharing control is valid in the sense that the sojourn time is proportional to the required service; a task requiring x s of service will stay on average $T(x) = x/(1-\rho)$ in the server (ρ denotes as usual the load of the server).

The state probabilities are known; they are in fact identical with those of the queue with processing in the order of arrivals, the discipline being conservative.

Reference should be made to [Kleinrock 75] for a more complete study and the application to time-shared data-processing systems.

9.4 Application of simulation

Simulation is often used for the study of the control system for similar reasons to those mentioned in Chapter 4.

- Analytical methods fail to deal realistically with a large number of specific operational characteristics (complex task scheduling, nonexponential service times, etc.) as the preceding sections have shown. Certain operating modes are quite simply inaccessible to analysis (cf. Section 9.3).
- It is a difficult requirement that analytical methods should predict certain parameters (quantiles of the preselection time).
- Only simulation seems suitable in the current state of our knowledge to study operation under overload conditions (and especially the mechanism of regulation; this involves observing transient phenomena) (see Chapter 10).
- Finally, simulation appears often as a means of attack which is direct and accessible to all, in contrast with the analytical methods.

We shall no longer speak of "simulation by roulette", the assumptions supporting it not being met here; it is therefore a matter of classical "real-time simulation". However, the cumbersome nature and the cost of a simulation experiment have made it necessary to look for other approaches.

Direct simulation consists in reproducing entirely the structure of the control system in the simulation and in describing the routing of each type of call; the faithful reproduction of routing makes it necessary to simulate the durations of dialing, ringing, conversation and signaling.

Thus an outgoing call in a local exchange gives rise to the sequence described in Fig. 9.6.

The total duration of the call is of the order of 3–5 min. The full squares correspond to effective interventions of the control system (15–50 ms for the processor, and 100–300 ms for the active units of the periphery).

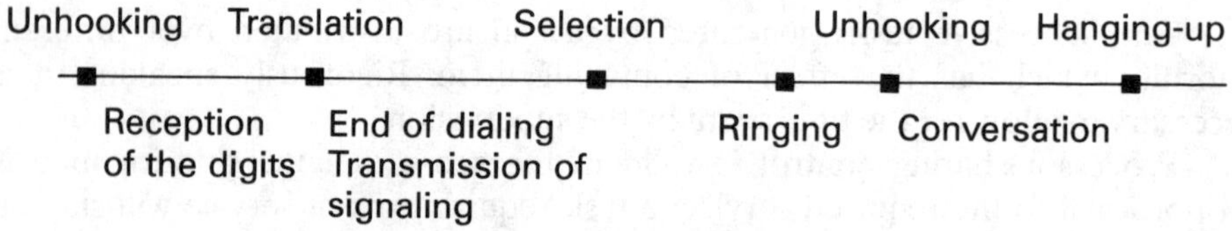

Fig. 9.6 Active periods in the simulation of the control system.

The life of the call in the simulation will therefore consist of point actions followed by long periods of staying in the future event set; during the periods between digits, during the conversation etc. the control system in effect totally ignores the call.

Simulation by call segments makes use of the difference in the time scales involved in order to simplify the description. Since a very long time (relatively) elapses between two successive passages of the call into the control system, the situations encountered by the call in its successive phases can be considered as independent. The faithful reproduction of the chaining with respect to time therefore makes no sense.

A stream of independent call segments is simulated. In the above-mentioned example, the initial stream is replaced by five independent streams passing through the model: preselection, translation etc. The parameters of these streams can be derived directly from the traffic model; thus, there are more "preselection" segments than "conversation" segments because of the abandonments and failures.

The simulation by call segments has a certain number of advantages over direct simulation.

- Each call occupies the simulator for approximately 3 min in direct simulation. This means that 3000–5000 future events permanently occupy the event file. In simulation by call segments, only about ten events have to be permanently controlled; this means an enormous gain in memory and especially in time for handling the future event file.
- The initial transient period, the "start-up", becomes very short; according to the experiments reported in [Dietrich 77], it is of the order of 10–20 s as against several tens of minutes in simulated time for direct simulation; this simplifies the removal of initialization bias.
- Finally, of course, the description is simplified since apart from the chaining, only the information belonging to each call undergoing progress needs to be stored.

These advantages pay off.

- The model has available common resources, in particular memories, which are occupied throughout the duration of the call. They play an important part in the detection of overload (for which they may be responsible). The simulation by call segments obviously cannot reproduce their occupation and cannot be used as a basis for their dimensioning.
- The assumption of independence between segments is certainly not verified during the congested period when the overload control acts. Thus the calls rejected during preselection do not cause selection. It is therefore necessary to construct an auxiliary model for calculating the flows of call segments.

We shall now give an alternative method which offers the advantages of the simulation by call segments but reproduces the correlation between phases;

starting from a direct simulation, it consists of eliminating the passive residence times in the memory. We return to the situation of the simulation by call segments but this time observing the determinism of the segment chaining. Nevertheless, the first disadvantage still remains.

Other approximations have been devised in order to reduce the effort involved in simulation (writing and operation) at the cost of exactness. Some studies have mixed simulation with the analytical model.

The type of approximation depends on the objective required. For example, it is sufficient to simulate the central processor and its scheduling mechanism by reproducing the load which is offered to it by priority level or job queue. The flows on the different levels are correctly represented in magnitude but are independent. An example of the application of such an approximation is found in [Itoh 76] where the precision is judged by comparison with direct simulation. The fit appears to be quite correct.

Contrary to the simulation of switching networks, the use of specialized languages often turns out to be very useful. The application of simulation by call segments permits reasonable sizes of future event sets; the combinatorial complexity of the described system is low and the control of the chaining represents a significant part of the activity, all circumstances favoring a specialized simulation language (SIMSCRIPT, GPSS etc.).

Finally, we indicate the possibility of developing a "universal simulator" for the control structures (see [Addie 80]).

However, simulation should not be regarded as an infallible tool for the traffic specialist; in this application as in others, it suffers from limitations, the main ones being as follows.

- The cost in writing and developing the program combined with the cost of operation which is high because of the expansion ratio.
- The practical impossibility of determining the precision of a measurement (confidence interval).

As an example, the simulation described in [Itoh 76] took up 24 man months for writing and 34 man months for development. The program (6000 instructions in FORTRAN and 2000 in Assembler) took up 80 000 words of memory.

The expansion ratio of the program, i.e. the central processing unit time consumed divided by the simulated time, varies from 10 to 30.

In the experiments described in [Dietrich 77], it is of the order of 1–2; the benefit gained from using simulation by call segments can be seen.

9.5 Conclusion

This discussion of the field of queuing theory applied to the study of exchange control systems has clearly shown the advantages and the limitations of the theory; it gives powerful synthetic results but has a limited area of application. Simulation can overcome these drawbacks but it is dangerous to use it in isolation. It seems to make sense to combine these methods. The method in Section 9.1 gives a general framework, and the overall performances are obtained by breaking down and analyzing each element in isolation. Mathematical results exist depending on the case and they are then used. If they do not exist, this will be the case when priority rules and clock interrupts are combined or with complex cyclic control. The element posing the problem can then be handled in the form of a simulation program which is necessarily of limited scope and therefore low in cost and rapid to implement; its efficiency (precision) will be tested by comparing it with the theoretical system resembling it the most (thus a program simulating the waiting process in a $GI/GI/1$ queue will be tested by comparison with known formulas in the case where $GI = M$).

Finally, in the design phase of the hardware, the precise rules which are unknown do not hinder the application of approximations; mathematical methods can be applied.

References

In this chapter, most of the results have been given without proof because of the mathematical techniques which would have to be employed. The references nevertheless fill the gap and provide extra information.

The queuing theoretical aspects of the chapter are well presented in [Kleinrock 75], [Gelenbe 80] and [Lavenberg 83].

Most of the references to Chapters 7 and 8 present more or less complete case studies.

[Addie 80] R. Addie, G. Codsi, P. Gerrand, M. Rubinstein. A general exchange simulator for capacity studies of telecommunication systems. *Australian Telecommunication Research*, Vol. 14, No. 1, 1980.

[Altamirano 79] M. Villen Altamirano, G. Morales. Traffic analysis of a class of distributed systems. *Proceedings of the 9th International Teletraffic Congress*, Torremolinos, 1979.

[Altamirano 83] M. Villen Altamirano. Average response times in an M/G/1 queue with general feedback and priorities. *Proceedings of the 10th International Teletraffic Congress*, Montreal, 1983, Paper 5.2.5.

[Bux 83] W. Bux, H. L. Truong. Token ring performance and mean delay approximations. *Proceedings of the 10th International Teletraffic Congress,* Montreal, 1983, Paper 3.1.3.

[Chan 73] W. C. Chan, W. K. Chung, D. Y. Maa. Multiserver computer controlled queuing system with preemptive priority and feedback. *Proceedings of the Institution of Electrical Engineers,* Vol. 120, No. 11, November 1973, pp. 1331–1336.

[Dietrich 77] G. Dietrich, R. Salade. Subcall type control simulator of SPC switching systems. *Electrical Communications*, Vol. 52, No. 1, 1977.

[Eisenberg 72] M. Eisenberg. Queues with periodic service and changeover time. *Operations Research*, Vol. 20, 1972, pp. 440–451.

[Enns 69] E. G. ENNS. Some waiting time distributions for queues with multiple feedback and priorities. *Operations Research*, Vol. 17, 1969, pp. 519–525.

[Enns 70] E. G. ENNS. A simple method of calculating queue size distributions of priority queueing systems with feedback. *Proceedings of the 6th International Teletraffic Congress*, 1970, Paper 314.

[Fujiki 76] M. FUJIKI, Y. MURAO. Queueing models with regular service interruptions. *Proceedings of the 8th International Teletraffic Congress*, Melbourne, 1976.

[Gay 75] T. W. GAY, P. H. SEAMAN. Composite priority queues. *IBM Journal of Research and Development*, January 1975, pp. 78–81.

[Gelenbe 80] E. GELENBE, I. MITRANI. *Analysis and Synthesis of Computer Systems*, Academic Press, New York, 1980.

[Gerrand 76] P. GERRAND, A. GUERRERO. The waiting time distribution for markers and other peripheral devices in SPC switching systems. *Proceedings of the 8th International Teletraffic Congress*, Melbourne, 1976, Paper 613.

[Hakansson 70] L. HAKANSSON. A GPSS simulation study of the capacity of a SPC exchange computer. *Proceedings of the 6th International Teletraffic Congress*, 1970, Paper 425.

[Hashida 72] O. HASHIDA. Analysis of muliqueues. *Review of the Electrical Communication Laboratories*, Vol. 20, 1972, pp. 189–199.

[Haugen 76] R. B. HAUGEN, E. JENSEN, M. J. SANCHEZ-PUGA. Analytical study of feedback effects on processor traffic in SPC. *Proceedings of the 8th International Teletraffic Congress*, Melbourne, 1976, Paper 624.

[Herzog 75] U. HERZOG. Optimal scheduling strategies for real time computers. *IBM Journal of Research and Development*, Vol. 19, September 1975.

[Herzog 76] U. HERZOG. Priority models for communication processors. *Proceedings of the 8th International Teletraffic Congress*, Melbourne, 1976, Paper 623.

[Itoh 76] M. ITOH, T. UEDA, K. OKADA. Methods of estimating central processing system traffic performance in SPC switching systems. *Proceedings of the 8th International Teletraffic Congress*, Melbourne, 1976, Paper 621.

[Jaiswal 68] N. K. JAISWAL. *Priority Queues*, Academic Press, New York, 1968.

[Jung 81] M. JUNG, J. DE BOER. Waiting time distribution for call processing tasks in a SPC telephone system, offered via a clock-pulse operated gate. *Philips Telecommunication Review*, Vol. 39, No. 2, June 1981, pp. 90–101.

[Kleinrock 75] L. KLEINROCK. *Queueing Systems* (2 volumes), John Wiley, New York, 1975.

[Kuehn 76] P. J. KUEHN. Analysis of complex queueing networks by decomposition. *Proceedings of the 8th International Teletraffic Congress*, Melbourne, 1976, Paper 236.

[Kuehn 79] P. J. KUEHN. Multiqueue system with nonexhaustive cyclic service. *Bell System Technical Journal*, Vol. 58, No. 3, March 1979, pp. 671–698.

[Labetoulle 80] J. LABETOULLE, G. PUJOLLE. Networks of queues. *IEEE Transactions on Software Engineering*, Vol. SE6, No. 4, 1980, pp. 373–381.

[Lavenberg 83] S. LAVENBERG. *Computer Performance Evaluation Handbook*, Academic Press, New York, 1983.

Chapter 10

REATTEMPTS; OVERLOAD

In a telecommunications system (exchange or network), as in any real-time system, the overload is of critical importance. Overload is understood as a significant increase in demand (offered traffic) above the nominal value. The effects of overload are varied and can go so far as a collapse in the performances, making the carried traffic fall to zero whereas the offered traffic grows immeasurably.

These dramatic effects are due to the behavior of the users and in the first instance are due to reattempts. It can be seen that this phenomenon is important and the value of having available a model, even simplified, of the behavior of users faced with overload conditions is obvious.

Section 10.1 presents the main phenomena which are involved: impatience, repeated calls and some typical values for typical behavior parameters of subscribers.

Section 10.2 presents a theory of reattempts which allows a global representation and a simple characterization of the phenomenon. Its application largely goes beyond the scope of exchanges; in fact, it is the area of the network (interexchange) which constitutes its main field of application. This is a "macroscopic" model relating the global variables to each other. It has to be completed by a specific "microscopic" model which depends on the phenomena studied: transit network, PABX (behavior of the operators) or exchange.

Section 10.3 shows how the behavior of the subscribers (impatience, faulty operations and of course reattempts) will modify the operation of the exchange and cause and then aggravate the overload; this is an avalanche effect whose origin may be a true overload or an accident (fault or equipment out of service).

The phenomenon takes different forms according to the architecture of the system. All structures are not sensitive to overload to the same extent. Prevention or regulation methods are effective in combating its harmful effect. This is the overload control system, some versions of which are described in Section 10.4 as well as the problems of evaluation which they present.

10.1 User behavior

10.1.1 Impatience

Under fluid traffic conditions, the users experience their calls being made quickly and well (imperceptible waiting for the tone, rapid selection etc.).

As soon as the load increases, the response times become longer and, to begin with, the reception of the dialing tone (preselection phase). Two reactions of impatience can thus be observed.

- Premature dialing. The subscriber, who is used to a tone which is nearly always instantaneous, starts his dialing without checking the presence of any tone. The first digits are therefore altered or lost. Most often, when the fixed format of dialing in use is borne in mind, the exchange will record too few numbers, the call will remain on standby and will be rejected after the elapse of a time delay. From the user viewpoint, an unsuccessful attempt results; from the exchange viewpoint, the control units will have worked to no avail (occupation of circuits, and processor time).
- Premature hanging-up. The subscriber who does not hear the normal signals hangs up whereas his call is perhaps being processed. The "patience" time varies with the reaction of the exchange, the phase of the call and the destination. The importance of this type of impatient behavior encourages the use of tones which "reassure" the user. The tone which is used on the French trunk network explains for instance the difference in the user reaction during national calls compared with international calls. For more details, reference should be made to [Roberts 81].

Irrespective of the type of impatient reaction, its effect on the control system of the exchange and also on the telephone network is the same, i.e. inefficient use of resources; the computer assigns some of the processor time, some of its memory etc. to priority but useless tasks and this at the time when these resources are most precious.

Furthermore, all these inefficient attempts contribute to the reattempt traffic.

A campaign already out of date is reported in [Wilkinson 68]; it concerns users in the USA. More recently, Burkard *et al.* have performed the same type of measurements. Some of their results are summarized in Fig. 10.1 [Burkard 83].

In [Roberts 79] the distribution of patience for trunk and international calls is given (Fig. 10.2); this is the time elapsing between the end of dialing and abandoning the call.

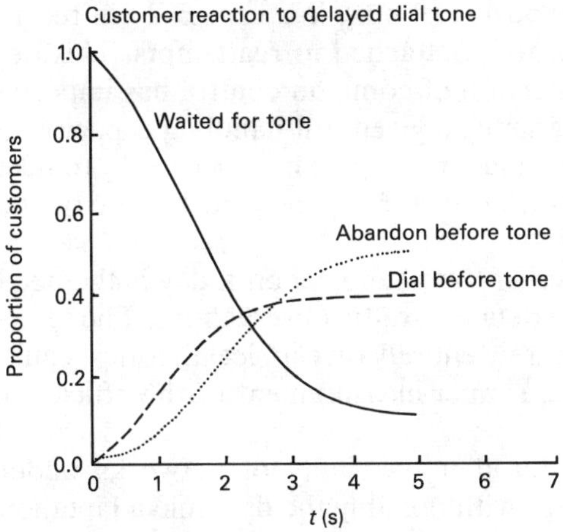

Fig. 10.1 Impatience during preselection.

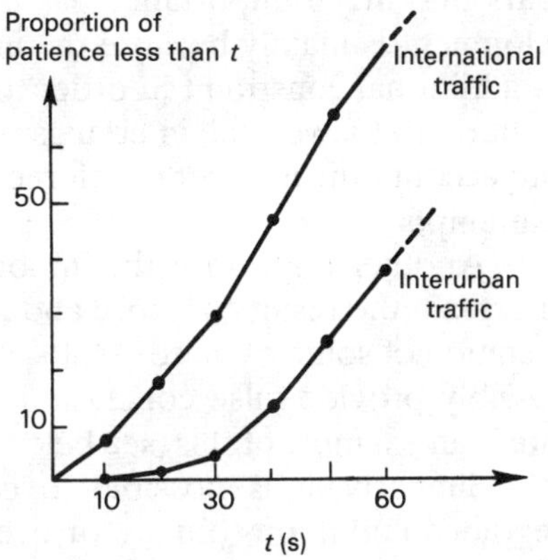

Fig. 10.2 Impatience during selection.

10.1.2 Reattempts

The traffic formulas usually used for designing telecommunication networks only allow for certain characteristics of subscriber behavior. Thus, classical queuing models assume potentially indefinite patience of the customers in the queue; they also ignore any idea of "wrong operation". We have already discussed, in the previous section, various causes of abandonment not predicted by the classical models and their consequences on the control system.

The Erlang loss formula assumes that the call arrival process obeys a Poisson distribution and that the blocked calls leave the system finally. Our own experience also teaches us that the subscriber who receives the busy tone will in most cases repeat his call after a more or less brief interval.

All these examples show up the inadequacy of the simplified model; there will be a larger number of abandonments or rejections than predicted and the resulting reattempts will markedly modify the nature of the offered traffic. What are the consequences of these reattempts on the quality of the telephone service?

In the "classical" approach, the offered traffic is a fundamental parameter and it can be equally regarded as the demand which would be satisfied if an infinite quantity of resources were available or as the number of call arrivals during the average occupation time (definition in the sense of the Erlang formula). If, however, reattempts are taken into consideration, it is necessary to accept that the effective arrival rate is in fact a function of the blocking and cannot serve alone as a characterization of demand.

Although the phenomenon is obviously not new, it is especially in recent years that much importance has begun to be attached to reattempts. This development is mainly because the introduction of common control has imposed an additional constraint in order to guarantee a given call-handling capacity; in addition to the volume in erlangs of the voice traffic, it is now necessary to take into account the number of offered calls. This factor depends much more on the reattempts.

Another reason for the importance of the phenomenon today is the need to explain the results of more and more detailed traffic observations. The interpretation of some of these results, which rely entirely on classical theories, could possibly provide false conclusions. The simple measurement of the efficiency rate is an example of this (see below).

Similarly, it is possible to consider that the study of networks under degraded conditions (failure or overload) with possibly the dynamic adaptation of the traffic routing as a function of the observed grade of service must take into account the strong interaction between the behavior of the network and that of the subscribers expressed by the reattempt phenomenon.

The study of reattempts makes it necessary to define the terminology. The definitions which follow are taken from the resolutions of the Terminology Group of Commission II of the CCITT.

- Call attempt is the operation by which the caller attempts to establish a telephone communication. At any point in the routing of a communication, only attempts having reached this point are taken into account. An attempt is translated by an established call (one then speaks of an effective attempt), or by a failure and then a reattempt may occur.
- Call demand is the desire to etablish a communication which is manifested by one or more successive call attempts. It is also called a call intention. Only call attempts defined above may be taken into consideration.
- Fresh call is the first call attempt observed for a given call demand. The subsequent calls are "repeated calls" or reattempts.
- Call repetition coefficient β is the average number of call attempts per call intention.
- Efficiency rate (completion ratio) r is the ratio of the number of effective call attempts to the total number of call attempts.

In order to measure the phenomenon of call attempts and to obtain a better understanding of its consequences on the telephone network, several observation campaigns on subscriber behavior have been undertaken all over the world. These require the mobilization of considerable resources for the acquisition of data as well as for their subsequent analysis.

Despite the fact that most observation campaigns have taken place in the telephone network and not inside an exchange, they reflect the user's reaction and are completely usable within the framework of the exchange. This is all the more true when most of the time the user cannot detect the causes of his failure with any precision.

The observation and analysis procedures are complex and require extreme care [Roberts 79].

The user's reaction (to a failure) is characterized by the probability of perseverence and the time before a new attempt.

The reaction of the network is characterized by the failure rate (and the cause) which is a function of the rank of the attempt and the type of call.

Probability of failure

Table 10.1 shows the relative variation in the causes of failure (in the network) with the rank of the attempt. The values taken from [Roberts 79] originate from a sample of more than 350 000 calls.

Table 10.1 Percentage of the various "fates" of call attempts

Number of the attempt	*Local calls*				*Interurban calls*			
	Effective	*Busy*	*No reply*	*Miscellaneous*	*Effective*	*Busy*	*No reply*	*Miscellaneous*
1	77	9	9	5	72	11	6	11
2	48	28	15	9	49	25	9	17
3	36	38	16	10	41	31	9	19
4	31	45	14	10	35	36	8	19
5	25	52	13	10	30	39	9	22
6	23	54	13	10	32	38	8	22
7	22	57	12	9	27	40	9	24
8	19	61	11	9	23	45	7	25

Each row gives the percentages relative to the number of calls of the rank concerned. Thus, 31% of the calls issuing a fourth attempt are completed and 45% result in a busy tone, etc.

It should be noted that the probability of failure (in particular due to occupation) increases markedly with increasing rank of the attempt. Two main reasons have been identified to explain this increase.

(1) A reattempt soon after a failure will very probably find the same cause of failure.

(2) Reattempts have a higher probability of failure because the fact of failing at the first attempt selects a sample of call intentions whose destination has a lower efficiency than average.

The increase in this failure probability explains why the efficiency is less than would be predicted by a theory of lost calls. This would be more suitable for calculating the failure rate of the first attempts.

Perseverance probabilities

Table 10.2 shows the percentage of calls repeated during the day as a function of the rank of the attempt and the cause of failure.

Table 10.2 Percentages of reattempts

Number of the attempt	*Urban calls*		*Interurban calls*	
	Busy	*No reply*	*Busy*	*No reply*
1	81	55	90	58
2	86	61	92	62
3	88	67	93	71
4	90	70	94	74
5	90	71	94	69
6	91	76	97	75
7	92	81	93	80
8	92	75	95	70

Perseverance is especially high when the subscriber called is busy. It grows with the rank of the attempt which can be explained by progressive selection of subscribers who are *a priori* the most persevering out of the population of observed subscribers.

The curves in Fig. 10.3 show the distribution of the intervals between the first and second attempts.

The intervals are shorter if the failure is due to a subscriber who is busy.

10.2 A theory of reattempts

There is no theory of reattempts which is universally accepted and applied. We shall describe below the basic principles of a very general descriptive model well suited to the measurements described in the previous section.

10.2.1 The global model

The theory is split into two stages. First the "global model" calculates the macroscopic behavior of the system and the value and composition of the total offered traffic as a function of the parameters of subscriber behavior and the failure probabilities. Then, each specific study will require a specific model which relates the probabilities to the offered flows and to the detailed operation. Here, we shall consider the global model.

In reality, the global model provides a general method for interpreting the observed phenomena. H is the probability of perseverance; depending on the model, it depends on the cause of failure or the rank of the attempt. p is the probability of rejection (by the system, whatever it is) which is also a function of the cause or of the rank. λ_0 is the fresh offered traffic, λ_1 the total traffic (fresh

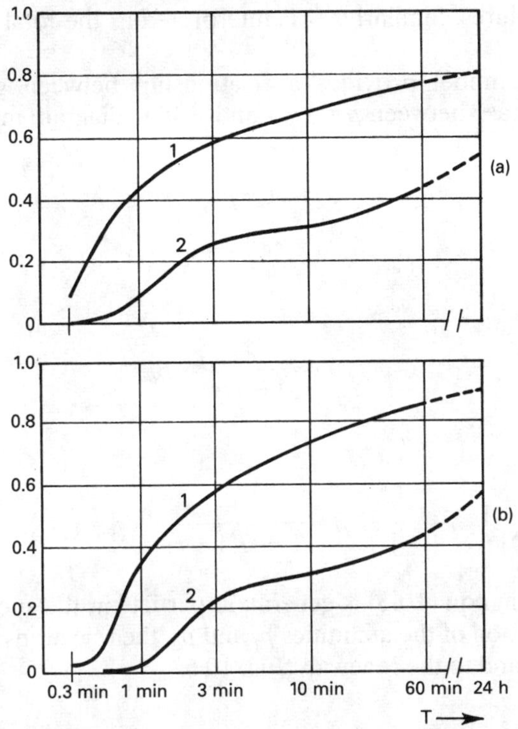

Fig. 10.3 Probability that a second attempt is observed depending on the cause of failure and that the interval between the first and second attempts is less than T for (a) local calls and (b) national calls: curves 1, subscriber busy; curves 2, no reply.

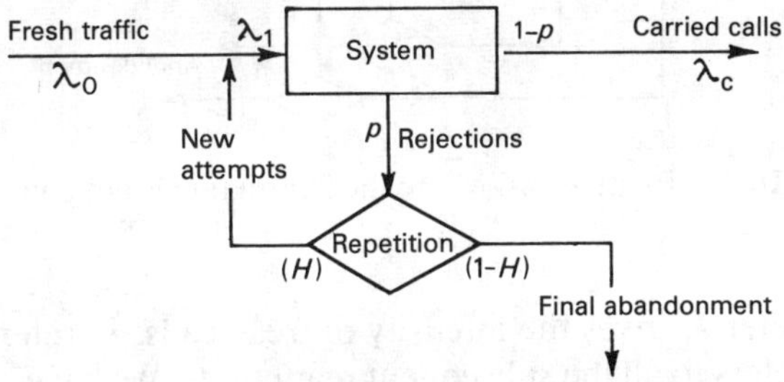

Fig. 10.4 Model of traffic with reattempts.

traffic and subsequent attempts) and λ_c the carried traffic (which terminates in conversation).

The diagram in Fig. 10.4 defines the elementary reattempt model.

In this simple system, no attempt is made to identify the causes of rejection; the probability p integrates all the sources of rejection and models the observed proportion of failures. Similarly, H is interpreted as the total proportion of reattempts.

The whole model provides a relationship between global observable variables, in our case between $p = 1 - r$ and β. The diagram makes it possible to write directly

$$\begin{aligned} \lambda_c &= (1-p)\lambda_1 \\ \lambda_1 &= \lambda_0 + Hp\lambda_1 \end{aligned} \tag{10.1}$$

and

$$\lambda_1 = \frac{\lambda_0}{1 - Hp} = \beta\lambda_0 \tag{10.2}$$

i.e.

$$\beta = \frac{1}{1 - H(1-r)} \tag{10.3}$$

The relationship in eqn (10.3) is general; in particular, the specific model often leads to the definition of the quantities p_1 and p_2, the rejections corresponding to the first attempts and to the renewals (Fig. 10.5).

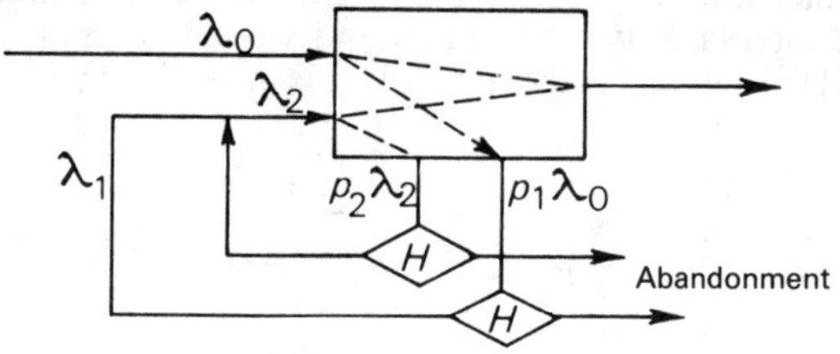

Fig. 10.5 Reattempts, where the failure rate is not constant.

The parameter λ_0 gives the intensity of fresh calls, λ_1 refers to the first reattempt and λ_2 refers to all the subsequent reattempts. We have

$$\begin{aligned} \lambda_1 &= p_1 H\lambda_0 \\ \lambda_2 &= \lambda_1 + p_2 H\lambda_2 \end{aligned}$$

i.e.

$$\lambda_2 = \frac{p_1 H}{1 - Hp_2}\lambda$$

The global coefficients are defined as above:

$$r = \frac{\text{carried traffic}}{\text{total observed traffic}}$$

$$= 1 - \frac{p_1}{1 - H(p_2 - p_1)}$$

$$\beta = \frac{\text{total observed traffic}}{\text{fresh traffic}}$$

$$= \frac{1 - H(p_2 - p_1)}{1 - Hp_2} = \frac{1}{1 - H(1 - r)}$$

The value of the reattempt model can be seen here; the efficiency rate, which is the measured variable, depends on the characteristics of the installed system via the parameters p_1 and p_2 but also on the subscriber via the factor H.

As we have already stated, the model (eqns (10.1) and (10.3)) should be seen as a general framework which can be adapted to the situations under study. Thus a call has to overcome several obstacles (passage through its originating exchange, passage through one or more local exchanges and the presence or absence of the called party). A different pair (p, H) corresponds to each stage as has been shown by the measurements, assuming that the parameter p is not a function of the rank of the attempts but only of the stage which has been reached and H constant (Fig. 10.6).

It should be noted that here the index of p refers to the obstacle and not to the rank.

The cascade of obstacles may always be represented by a single probability; in the example in Fig. 10.6, the reader can easily verify that

$$\beta = \frac{\lambda_1}{\lambda_0} = \{1 - (1 - r)\mathrm{H}\}^{-1}$$

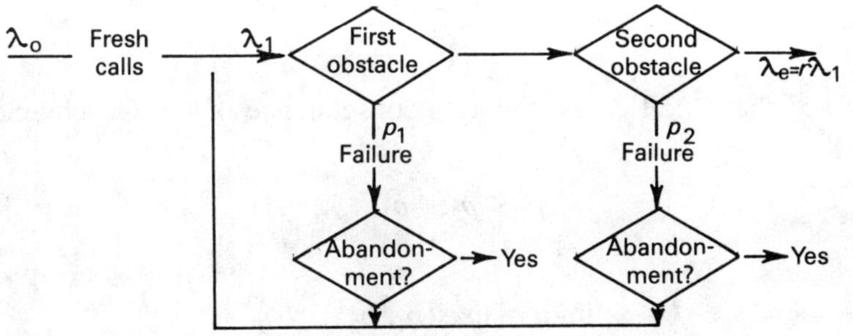

Fig. 10.6 Cascading obstacles and reattempts.

where

$$r = 1 - \{p_1 + (1 - p_1)p_2\}$$

The model thus leads to the idea of the global "efficiency rate of the network", a resultant of combining the failures. Similarly, from the viewpoint of the caller subscriber, an efficiency rate encompassing the called party is defined.

The global model also makes it possible to explain the growth in failure probabilities observed with the rank of the attempt.

Let us observe an exchange which serves n outgoing directions (trunk groups) with failure rates b_i, $i = 1, \ldots, n$. Observation provides a global failure rate which is the mean value over all directions while distinguishing between the ranks of the attempts. Let a_i be the (fresh) traffic offered to the direction i.

The proportion of failures at the first attempt is

$$p_1 = \frac{\sum_i a_i b_i}{\sum_i a_i}$$

The result is $\sum_i a_i b_i H$ second attempts, of which

$$p_2 = \frac{\sum_i a_i b_i H b_i}{\sum_i a_i b_i H}$$

fail. In general, the failure rate at rank k will be

$$p_k = \frac{\sum_i a_i b_i^{k}}{\sum_i a_i b_i^{k-1}}$$

We then see that in all cases (this is a consequence of the Cauchy–Schwartz inequality)

$$p_1 < p_2 < p_3 < \ldots$$

and

$$\lim_{j \to \infty} p_j = \max(b_1, b_2, \ldots, b_n)$$

10.2.2 Specific model

The specific models make it possible to estimate the p_i. There will be as many specific models as systems studied. The case of a trunk group (or more generally of a group of servers) has been intensively studied.

An immediate model [Kosten 47] consists in estimating the failure probabilities by means of time-dependent solutions of the state equations. The probability of failure of a first attempt is the steady-state solution since it can be assumed that the call intentions arrives randomly (Poissonian traffic). However, for a reattempt, if $F(t)$ denotes the distribution of the time between attempts,

$$p_2 = p_3 = \ldots = \int_0^\infty P(\text{all servers busy at } t \mid \text{all busy at } 0)\,\mathrm{d}F(t)$$

Other more elaborate models have been suggested (see [Kosten 47], [Cohen 57], [Elldin 67] and [Legall 77]).

Most of the models studied fall within the Markovian framework; an exponential distribution is assumed for the duration between successive attempts and the reattempt process is then characterized by the total number of calls awaiting reattempts (see [Wilkinson 68] and [Legall 77] for the case of an isolated trunk group).

The approach used in the next section (eqn (10.6)) consists in replacing the flow of all attempts by an equivalent flow with a parameter $\lambda_1 = \beta\lambda_0$. The system equation provides a relationship between the rejection and λ_1 whose elimination yields the solution. This is a first-order model which assumes that it is possible to express the rejection by means of a conservation equation. Care should be taken that in the general case the resultant traffic (here λ_1) is not at all Poissonian.

10.3 Overload in exchanges

10.3.1 Description of the phenomenon

Up to this point, the control system has been studied assuming the existence of steady-state conditions; at the end of a long time, all the observable variables (the establishment time) fluctuate around finite average values and all the customers end up by being serviced.

The existence of this long-term equilibrium implies that the utilization rate of each resource (circuits, memories, connection bus and above all the processors) remains moderate (less than 0.7). The study carried out here intends to answer the question of what would happen if the load increases too much so that the utilization rates exceed this limit and tend to 1. This is the overload situation. In fact, it is necessary to have two concepts of overload in order to qualify the behavior of stored program control systems correctly.

Obviously, the overload is defined with respect to the nominal conditions. We start by fixing a certain number of indicators for the grade of service experienced by the user and a level for these indicators. The offered traffic which satisfies these criteria is the nominal traffic. For example, the preselection time is a good indicator; one of the specifications which give the nominal conditions in the CCITT standards is

$$P(\text{preselection time} \leqslant T_1) > P_1 \tag{10.4}$$

with $T_1 = 0.5$ s and $P_1 = 0.95$; these values ensure that under nominal conditions the users do not experience any trouble and do not disturb the operation.

What happens then if the offered traffic exceeds the nominal value? We know that it is possible to represent the exchange as a queuing network where servers are complex mechanisms. The response time is the sum of the time spent waiting and the time in service in front of each server visited. When the traffic increases above the nominal value defined by eqn (10.4), the waiting times increase and therefore also the preselection time. Under these conditions, the operation of the system remains identical with that under the nominal conditions but the grade of service is degraded. The standards define an overload traffic and an associated grade of service. The preselection time is a function of the load and the overload traffic must satisfy

$$P(\text{preselection time} \leqslant T_2) > P_2 \tag{10.5}$$

with usually $T_1 < T_2$ and $P_1 < P_2$. This is a necessary standard. It guarantees stable operation around the nominal point, protecting the system against any inaccuracy in the prediction of traffic, against day-to-day variations and against small but frequent overloads.

In fact, the juxtaposition of the constraints (10.4) and (10.5) defines the nominal conditions and an overload zone which can be called simple overload; in the French standards the simple overload traffic must be at least more than 40% of the nominal traffic. From the viewpoint of traffic flow study the methods described for nominal conditions apply perfectly to simple overload.

Nevertheless, if the traffic increases even more, it reaches a value where the occupation level of one of the resources (processor time, memory space, a peripheral etc.) reaches 100%. The waiting time increases inordinately since we are outside the "stability conditions" (cf. Chapter 2). Then the users hang up and renew, increasing the offered load. We speak of serious or catastrophic overload.

The "linear model" (Chapter 8) predicts for the occupation level of the "most heavily loaded resource" an evolution as shown in Fig. 10.7.

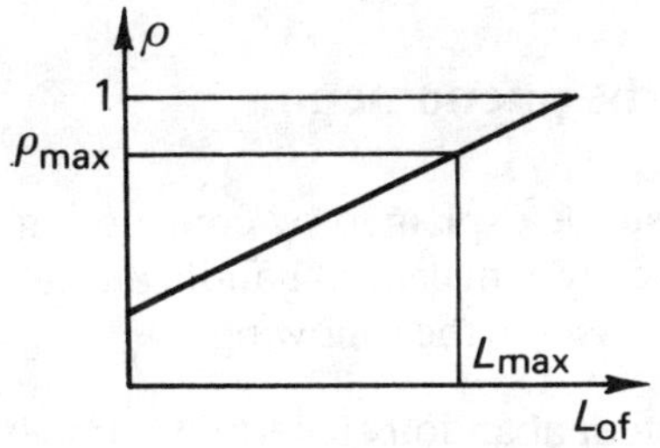

Fig. 10.7 Occupation level as a function of the offered traffic.

In the linear section of the curve, we have

$$\rho = \rho_0 + \bar{t}L_{of} \tag{10.6}$$

Applied to the processor of a centralized control system, ρ represents the control unit load, ρ_0 the background work, $\bar{t}$ the average processor time used in a call and L_{of} the offered traffic in number of calls per second. The extension can be easily made to any resource (Fig. 10.9).

If $\rho < \rho_{max}$, all the calls will be carried and therefore

$$L_{of} = L_c = \frac{(\rho - \rho_0)}{\bar{t}}$$

The value ρ_{max} which determines L_{max} is chosen strictly to be less than 1 in order to leave a reserve for the background tasks which cannot be delayed indefinitely. In addition, any waiting system exhibits unstable behavior if its occupation level is equal to 1.

Under serious overload conditions, i.e. above L_{max}, L_c must be less than L_{of}. The control system will not be able to handle all the calls and the user behavior in the face of an inordinate wait or a rejection will amplify the phenomenon as has been seen in the previous section; unsatisfied demands are repeated and congest the system whose response curve takes the form of Fig. 10.8 (if no precaution is taken).

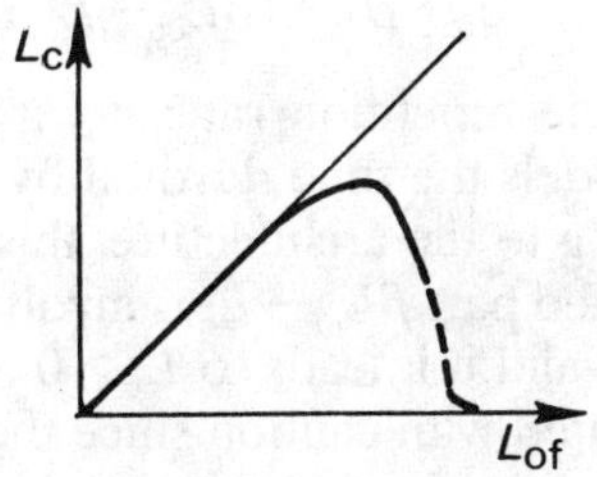

Fig. 10.8 Offered traffic and carried traffic.

10.3.2 A model of the phenomenon

The shape of the curve can be explained by considering the service mechanism. The calls which the control system cannot handle do not disappear; according to the architecture, we shall observe the following.

(a) The control system abandons a certain number of calls (usually in the course of preselection). Even if the abandonment is fast, the process of detection and the rejection test take a nonzero time (let us say 100 μs). The abandonment is invisible to the subscriber; the call will be detected again and may be rejected in the next scanning cycle. The system functions as a system with delay in which the users display impatience and abandon after a few seconds. Nevertheless the processor responsible for their processing will have detected, processed and rejected them a few dozens of times before they give up.

(b) The control system warns the user of its state of congestion (busy tone). The subscriber hangs up and perseveres by renewing his attempt a few seconds later. Here also, the rejection has taken up nonzero processor time.

(c) The control system accepts all the calls and undertakes to process them. Significant waiting time causes abandonment. We are in the same situation as in (a) or (b) but abandonment occurs later and hence a more considerable utilization of resources.

(d) Etc. (the reader can easily imagine other organizations and other rules).

These outline explanations are valid irrespective of the structure (centralized or distributed) if it is poorly protected. However, in a distributed structure the resource which forms the bottleneck may no longer be a processor. The interconnection bus, for example, may be this bottleneck. It is even possible to find several bottlenecks at the same time.

We note that calls have to be detected even under overload conditions in order to be able to give priority to so-called essential lines.

If we remain at the general level of a linear model applied to the most exposed resource, the above considerations will modify eqn (10.6) as follows ($L_{\rm of}$ refers to intentions and not to the observed total traffic):

$$1 = \rho_0 + \bar{t}L_{\rm c} + (\beta L_{\rm of} - L_{\rm c})\tau \qquad (10.7)$$

The parameter β is the repetition rate; it expresses the behavior of the users. The parameter τ models the time devoted by the vulnerable resource to an ineffective call (according to the architecture, this time will be a rather large fraction of $\bar{t}$); only the rejected part $\beta L_{\rm of} - L_{\rm c}$ is involved.

The equation is only valid if it leads to $L_{\rm c} > 0$ (for very high overload, the linear model should be treated with caution since the parameters will definitely vary in it).

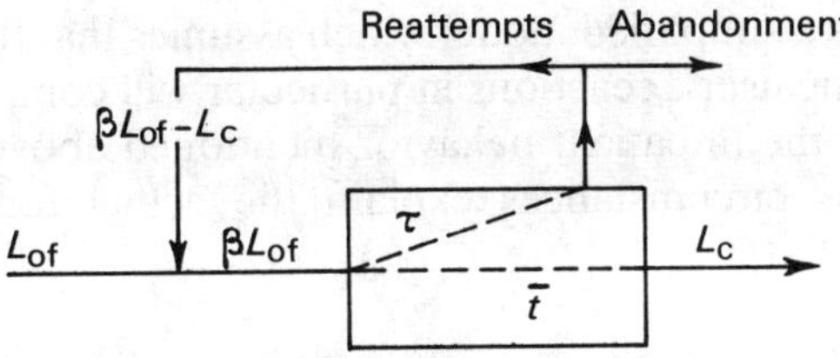

Fig. 10.9 Deterministic model of the overloaded exchange.

The reattempt model relates β to r, the system characteristic. Here,

$$r = \frac{L_c}{\beta L_{of}}$$

It is given by eqn (10.7). If this is combined with eqn (10.3), we obtain after a few manipulations

$$(1-\rho_0)(1-H) = L_c\{\bar{t}(1-H)-\tau\} + \tau L_{of} \tag{10.8}$$

This equation determines L_c as a function of L_{of}. We have illustrated this dependence in Fig. 10.10.

The relationship will only make sense for

$$\bar{t}(1-H)-\tau > 0$$

which corresponds to the fact that rejection takes up less time than efficient processing. If this is not the case, the model no longer applies; there are no steady-state conditions. In practice, this means that τ varies; rejection occurs earlier and earlier. We then observe curve (3) of Fig. 10.10 which also corresponds to the case where the users all have an infinite patience ($H = 1$).

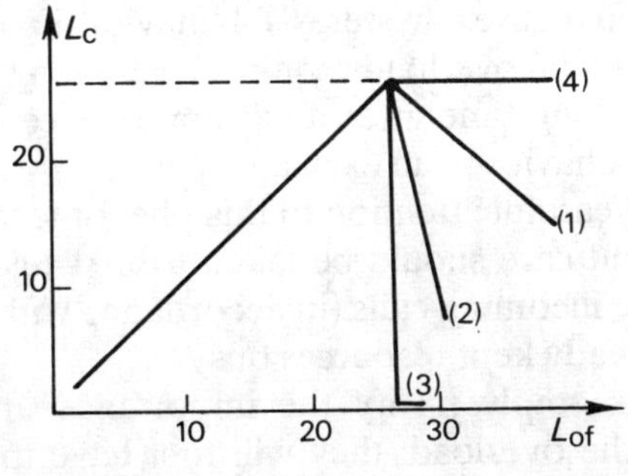

Fig. 10.10 Behavior on overload: curve (1), $\tau = 2$ ms; curve (2), $\tau = 5$ ms; curve (3), $H = 1$ with $\bar{t} = 30$ ms, $\rho_0 = 0.2$, $H = 0.8$; curve (4), ideal curve.

Of course, this is a simplified model which assumes that the parameters ρ_0, $\bar{t}$, etc. are constant. The user's reactions in particular will complicate the phenomenon. In particular, the impatient behavior mentioned above will take varied forms. These various circumstances explain the actual form taken by the response curve (Fig. 10.8).

10.4 Overload control

10.4.1 Consequences of the phenomenon

Congestion has been "explained" here on the one hand by the accumulation of calls not completely processed and which by virtue of their presence in the control system continually take up a percentage of the resources and on the other hand by the amplification effect of the users' reactions. Other systems require more detailed explanations at the "microscopic" level but the macroscopic effects remain identical and the above-described phenomenon is inevitable as soon as the offered traffic exceeds the value L_{max} if special precautions are not taken. The ideal system whose behavior conforms to curve (4) of course does not exist.

Since the exchange is part of the telecommunication network, it will react to all perturbations appearing at any point. Thus a reduction in the outgoing carried traffic (due to a transit exchange, say) will produce an increase in the call attempts originating in the exchange under study. However, a reduction in the incoming carried traffic due to local perturbation will cause a reduction in the efficiency rate of the network and an increase in the number of remote call attempts with the destination of the exchange (cf. Section 10.1). These, in their turn, cause blocking in the circuits and congestion in the transit exchanges. There is thus propagation and amplification of the overload and resulting inefficiency.

This amplification causes hysteresis behavior in the network and the exchange; a slight accidental overload requires a very long time to disappear and its effects persist for a long time after its disappearance (see [Forys 83] for an example of hysteresis behavior in an exchange).

It is worth paying careful attention to this phenomenon; it is inevitable that calls will be rejected but care should be taken to first reject outgoing calls and only as a last resort the incoming calls (in accordance with commonsense) since incoming calls have already kept resources busy.

All these reasons amply justify the importance attached to devices in charge of controlling the overload; they will first have to prevent their occurrence, then detect their appearance and finally reduce their effect.

Since, in the final analysis, the phenomenon is due to the increase in the offered traffic, the installed devices make the control system function as a queue (or a network) with limited capacity. The practical installation difficulties and the service constraints lead to more sophisticated implementations.

10.4.2 Regulation methods; centralized systems

Of course, it is preferable to avoid overload rather than to combat it; the most efficient method is passive filtering of the arrivals. In [Forys 83], there is a description of such a filter where a service of the type "last in, first out" avoids the annoying consequences of impatience. The equivalent diagram is shown in Fig. 10.11. The switch limits the number of customers placed in the buffer in a scanning cycle; the buffer is of limited size, and the new customers "push" the old customers who are ejected when there is overflow. The patient and disciplined users (who do not dial prematurely) are detected in each cycle and are put back in the queue; they increase their chances of being served. The server processes them (sending the tone and connection to a digit reception unit) by starting with the last in, which thus eliminates the risk of starting to serve a subscriber who will hang up or who has dialed too early. In [Forys 83] the performances of this control installed in the ESS1 system are evaluated and compared with an earlier method.

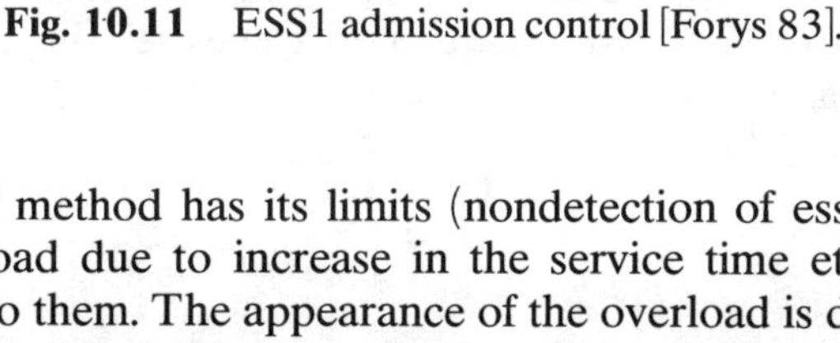

Fig. 10.11 ESS1 admission control [Forys 83].

This type of method has its limits (nondetection of essential lines, poor reaction to overload due to increase in the service time etc.); load controls should be added to them. The appearance of the overload is detected by observation of the occupation levels of a resource; the most simple to observe is the processor (the time spent on the background tasks or the cycle time for cyclic control). A large number of researchers have noted that the processor load due to call processing is not by itself a good indicator and prefer to observe several indicators (the offered traffic, if possible the response time of the calls or of certain phases, and the level of occupation of internal queues of the system) (see for example [Higuera 79]). In order to overcome fluctuations, the observations are smoothed out.

The performances which are expected of a detection are as follows.

- Rapid activation in order to avoid cumulative effects.
- Reliable detection which avoids premature activation; for this reason, smoothing is indispensable.

These two criteria which are clearly contradictory make compromise very awkward!

The protection methods consist in the last resort of cutting off the entry of new tasks. This is often done in a progressive manner as follows.

- The processing operations are partly removed by omitting noncritical parts of the call-processing software, e.g. preventing testing of equipment.
- The length of certain programs is then limited by limiting the number of restarts for unsuccessful attempts (search of equipment or paths); also the utilization of overflow trunk groups can be limited in order to avoid transferring the overload to the external network.
- The admission of new outgoing calls is limited; in the last resort, they are completely stopped.
- Finally, the admission of incoming calls is limited (or stopped).

The adopted scale respects the above-mentioned constraints; the first actions attenuate the impact of transitory overloads, and the incoming calls are only rejected as a last resort.

An example: the dynamic processor overload control (DPOC) method [Somoza 80]

This has been tested on the METACONTA 10C systems and is based on only the observation of the processor load. It consists in limiting the number of accepted calls during an interval of time (the "call-processing interval"). The limit number N is updated at the end of each "update interval" (Fig. 10.12).

The superscript (k) is used to refer to values relating to the last update interval. We calculate

$$\eta = n^{(k)} \frac{\rho_{\max} - \rho_0}{\rho^{(k)} - \rho_0}$$

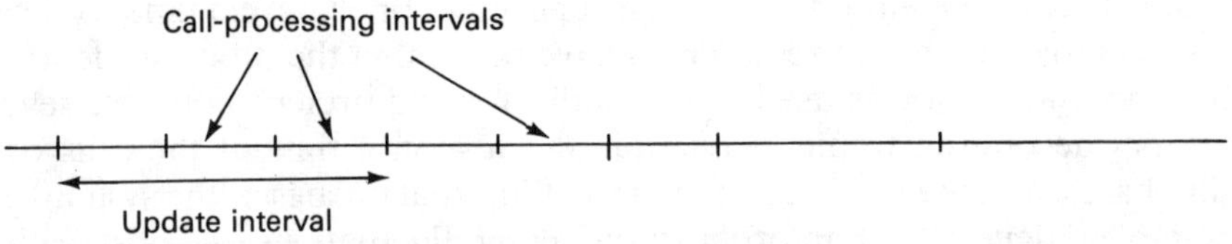

Fig. 10.12 DPOC method.

in which $\rho_{\max}$ is the maximum tolerated control unit occupation level, ρ_0 the empty control unit load, $\rho^{(k)}$ the last load measured (during the update interval) and $n^{(k)}$ the effective number of calls admitted during the update interval (k).

η will serve to calculate $N^{(k+1)}$. The relative variation is limited in order to avoid "oscillations" which would be caused without fail by a drastic variation in N:

$$\text{if } \eta > N^{(k-1)}, \text{ then } N^{(k)} = \min(\eta, N^{(k-1)}(1+e^+))$$

$$\text{if } \eta > N^{(k-1)}, \text{ then } N^{(k)} = \max(\eta, N^{(k-1)}(1-e^-))$$

where e^- (or e^+) is the maximum tolerated reduction (or increase).

Furthermore, in order to avoid an excessive load, the limitation is accelerated; if $\rho > \rho_M$, $N^{(k)}$ is multiplied by $(1 - \text{reduction})$.

Finally, if the occupation turns out to be too small, N is not modified.

For the algorithm to work in a suitable manner, it will need carefully chosen values for the time intervals and for e^+, e^- and ρ_M.

Satisfactory experiments (reported in the above-mentioned reference) used the following values: update interval, 4.5 s; processing interval, 1.5 s; $e^+ = 0.015$; $e^- = 0.1$; $\rho_{\max} = 0.95$; $\rho_M = 0.975$; reduction, 0.1.

Other methods have been devised; they can be found in the references.

As far as the evaluation of these regulation methods is concerned, all the studies given in the references proceed by simulation; we are dealing with a phenomenon with a very marked transient form under variable operating conditions. Analytical methods cannot reproduce the finite detail of the regulation mechanism and they often turn out to be incapable of handling such cases. Although it is the only solution, simulation should be handled with care, in this case also because of the transient aspect of the phenomenon. As already indicated in Chapter 8, environmental simulation is also of significant help in verifying the proper operation of the indicators and the setting of the thresholds.

Analytical models with limited use have been suggested (see for example [Schoute 83]). They provide a "qualitative" explanation of the phenomena or they serve to support and guide an exact analysis by simulation.

10.4.3 Regulation in distributed systems

In a distributed system and more generally in a multiprocessor system, the processor handling the calls is not the only unit sensitive to an overload nor the only one capable of detecting and regulating it. It is possible to delegate these tasks to specialized devices. This results in both a greater complexity of the controls to be carried out at multiple points and a greater flexibility and improved efficiency.

The main characteristic of the implemented procedures is to transfer as far as possible the waiting time and the congestion at the level of the "input units" (e.g. subscriber line switching units). These units control a small number of sources and so are not subject to a collapse in performance; the load on the control system is thus regulated and limited. This procedure guarantees an almost constant quality of service to the accepted calls (engaged in selection). Also the load on the control system which remains moderate guarantees the efficiency of the operating procedures, particularly in demand during overload or degraded conditions.

The large variety of architectures and the great interest in this field prevents us from making any final summary; we shall only review some implementations together with the problems of performance associated with them.

In [Doshi 83] a representative model of a large class of systems (Fig. 10.13) is analyzed. The peripheral processors are responsible for preselection and dialing reception and send a request for connection to the central processor; this calculates the routing, sends back the reply to the peripheral processor concerned which established the connection.

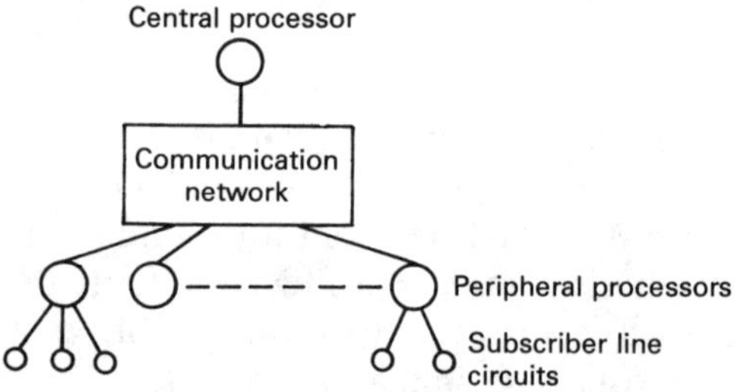

Fig. 10.13 A multiprocessor system.

This example illustrates the difficulty of control. It is easy for each peripheral processor to control its own state; if necessary a regulation system is installed there (similar to those of the previous section). The central processor supplied from several sources must limit its total traffic; how can the peripheral processor perform this limitation which ideally requires cooperation in order to establish a maximum flow and the smallest delays? In particular, it would be advantageous to handle imbalanced traffic (a heavily loaded peripheral processor is acceptable if the overall load remains moderate).

Let us assume nonoverloaded peripheral processors (there may be a large number of them, each then handling a small amount of traffic) and transparent to the traffic (except to the control); let λ_i be the flow offered by the peripheral processor *i*.

Doshi and Heffes [Doshi 83] suggest a series of control procedures inspired by the flow control methods in data transmission networks and which can be divided into two categories.

- Window control.
- Control of the speed of admission.

In the first type of control, each peripheral processor has K_i "tokens"; each message sent to the central processor travels on a token. In this way the number of requests which the peripheral processor *i* sends to the central processor is limited to K_i. A call arriving when no token is available will wait (or will be rejected depending on the size of the buffers). This control system can be represented by an equivalent circuit (Fig. 10.14). This is a closed network of queues with classes: K_i customers of class *i* with specific routing corresponding to the tokens of the processor *i*. The BCMP theorem is very suitable for this model.

In the second type of control the speed of sending the messages is limited. Each peripheral processor produces a token every *T* s (up to a maximum of *K* at the same time). As above, each message travels on a token (which here serves only once).

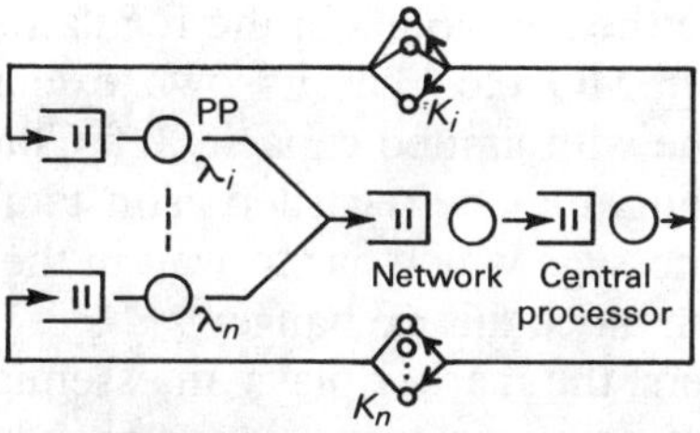

Fig. 10.14 Token control.

With these two strategies, Doshi and Heffes apply the service policy "last come, first served" with the same motivation as above (giving priority to the most recent calls which have the least risk of failing because of the user's impatience).

In the course of their analysis, they concluded that the second system is superior when it is installed in a dynamic version (the parameters *T* and *K* must be modified according to the load).

Although these approaches are interesting, it can be seen that they do not solve the problem of cooperation mentioned at the beginning of this section in a satisfactory manner. Each peripheral processor limits itself, the limitation being such that it allows the central processor to operate optimally only if all the peripheral processors offer identical loads; the case of imbalanced traffic is not solved.

Another interesting example of control is provided by the E10B system described in Chapter 7. We will briefly recall its structure (Fig. 10.15).

Control is effected at several levels: at the "local" level (line switching unit which filters on order from the control system), at the "filtering" level (the

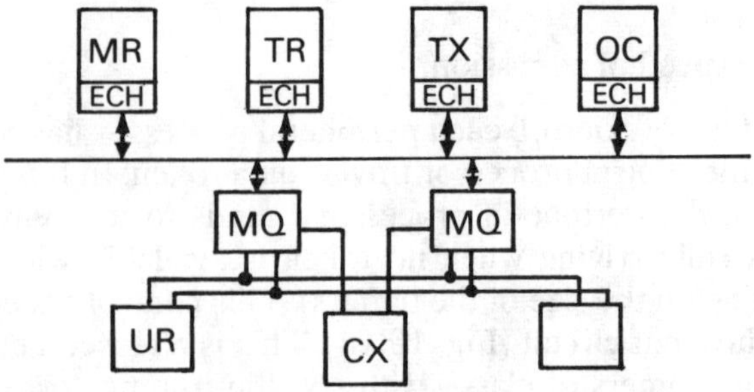

Fig. 10.15 The control system of the E10B system: MR, multiregister; TR, translator; TX, meter; MQ, marker; UR, line switching unit; CX, switching network.

control system accepting priority calls) and at the "central" level where the new calls no longer reach the control system.

The exchangers are the key points in the regulation. Each unit (the multiregister MR, the marker MQ etc.) has its own exchanger whose equivalent circuit diagram is a queue with limited capacity (cf. Chapter 7). The processing in the exchanger (switching) may be regarded as instantaneous; however, if it is full, it refuses any new message which has to wait in the exchanger of the transmitter (and may in its turn block this exchanger).

Of all the exchangers, the marker plays an essential role. In fact, this is an obligatory and a frequent point of passage in the processing (cf. Chapter 7, Fig. 7.12); its occupation level is directly related to the total number of calls being processed.

Furthermore, it is the marker which accepts new calls in the control system. An appropriate dimensioning will therefore limit the total number of calls in the control system which is the required objective.

"Deadlock", which is always possible with this type of structure (the exchanger A saturated by the messages to B saturated with messages for A), is overcome by means of time-out.

The dimensioning of the exchanger must guarantee a moderate occupation level to the unit which supplies it. The limited capacity of the queue is adapted to the more or less degraded state of the control units of the system; the bus makes it possible to know the state of all the units.

The evaluation of the performances of the architecture consists here in dimensioning the exchanger as a function of the required occupation levels. This is simply done by "ignoring" the overload aspect so that the model of evaluation with nominal load can be immediately applied. The advantage of the adopted architecture can be seen here.

Note that, apart from the evaluation of the operation of the control algorithm (as well as the setting of its parameters, thresholds or time constants), it may be desirable to estimate the overall effect of overload on the grade of service and on the network.

The estimation of the grade of service experienced requires the setting-up of an overall model of the exchange with the reattempt phenomenon added to it. The model currently used consists of a single queue with an exponential server submitted to a Poissonian flow of arrivals whose repetitions modify the arrival rate.

The operation of the exchange can be realistically allowed for by adopting a service level as a function of waiting time or of the number of customers [Forys 83].

The regulation mechanism must of course be incorporated in the model.

This is done by idealizing it; thus the queue will have limited capacity since the regulation consists in the final analysis of prohibiting access to the processor. The time spent by the regulation is taken into account by a variable service rate.

This thus reduces to a simple queuing problem.

The above-mentioned references provide an example of such an analysis which should not be overestimated, however; the regulation model is considerably simplified.

References

Reattempts have been subject to a large number of studies since the pioneering works of Cohen or Kosten (cited here mainly for their "historical" value); the works of Legall (most of them in French) have to be especially noted, since a large number of measurements and concepts have their origin here.

The observations reported in Section 10.1 are taken from [Burkard 83] and [Roberts 79]. See also [Myskja 76] and [Songhurst 83] for additional references.

For the modeling of impatience and an analytical study, see [Baccelli 81].

Some overload control methods are described in [Romoeuf 77], [Somoza 80], [Higuera 79], [Forys 83], [Fiche 81] and [Fiche 84]. In [Schoute 79] and [Schoute 83] an approach using optimal control is illustrated.

Finally, most system descriptions include a discussion of load regulation.

[Baccelli 81] F. Baccelli, G. Hebuterne. On queues with impatient customers. In *Performance 81* (ed. F. Kylstra), North-Holland, Amsterdam, 1981.

[Burkard 83] L. Burkard, J. J. Phelan, M. D. Weekly. Customer behavior and unexpected dial tone delay. *Proceedings of the 10th International Teletraffic Congress,* Montreal, 1983, Paper 2.4.5.

[Cohen 57] J. W. Cohen. Basic problems of telephone traffic theory and the influence of repeated calls. *Philips Telecommunication Journal,* Vol. 18, 1957.

[Doshi 83] B. T. Doshi, H. Heffes. Analysis of overload control schemes for a class of distributed switching machines. *Proceedings of the 10th International Teletraffic Congress,* Montreal, 1983, Paper 5.2.2.

[Elldin 67] A. Elldin. Approach to the theoretical description of repeated calls. *Ericsson Technics,* No. 3, 1967.

[Fiche 81] G. Fiche. Methods for performance evaluation of the E10B system (in French). *Commutation et Transmission,* No. 4, 1981, pp. 69–76.

[Fiche 84] G. Fiche, M. Ruoven. The E10 system: functional evolution and quality of service. *Proceedings of the International Switching Symposium,* Florence, May 1984, Paper 21A2.

[Forys 83] L. J. Forys. Performance analysis of a new overload strategy. *Proceedings of the 10th International Teletraffic Congress,* Montreal, 1983, Paper 5-2-4.

[Higuera 79] J. A. Higuera, C. D. Berzosa. Progressive method of overload control for processors. *Proceedings of the 9th International Teletraffic Congress,* Torremolinos, 1979.

[Kosten 47] L. Kosten. On the influence of repeated calls on the blocking probability (in German). *De Ingenieur,* Vol. 47E14, November 1947.

[Legall 77] P. Legall. The grade of service notion and repeated calls in telephony (in French). *Commutation et Electronique,* No. 59, October 1977.

[Myskja 76] A. Myskja, F. Aagesen. On the interaction between subscribers and telephone systems. *Proceedings of the 8th International Teletraffic Congress,* Melbourne, 1976.

[Roberts 79] J. W. Roberts. Recent observations of subscriber behaviour. *Proceedings of the 9th International Teletraffic Congress,* Torremolinos, 1979.

[Roberts 81] J. Roberts. Subscriber patience and post dialling delay for national and international long distance calls (in French). *Annales des Télécommunications,* Vol. 36, No. 11, 1981, pp. 595–601.

[Romoeuf 77] L. Romoeuf. A model of load regulation in a centralized stored program controlled exchange. In *Modelling and Performance Evaluation of Computer Systems,* North-Holland, Amsterdam, 1977.

[Schoute 79] F. C. Schoute. Optimal control and call acceptance in SPC exchanges. *Proceedings of the 9th International Teletraffic Congress,* Torremolinos, 1979.

[Schoute 83] F. C. Schoute. Adaptative overload control for a SPC exchange. *Proceedings of the 10th International Teletraffic Congress,* Montreal, 1983, Paper 5-2-3.

[Somoza 80] P. Somoza, A. Guerrero. Dynamic processor overload control for SPC switching systems. *Electrical Communications,* Vol. 55, No. 1, 1980, pp. 37–45.

[Songhurst 83] D. J. Songhurst. Subscriber repeat attempts, congestion and quality of service: a study based on network simulation. *Proceedings of the 10th International Teletraffic Congress,* Montreal, 1983, Paper 1.1.5.

[Wilkinson 68] R. J. Wilkinson, R. R. Radnik. Customers retrials in toll circuit operation. *Proceedings of the International Conference on Communication,* New York, 1968.

Chapter 11

THE EVOLUTION OF EXCHANGES WITH DISTRIBUTED ARCHITECTURES

The previous chapters have described traditional switching systems with SPC or rather the part of these systems sensitive to traffic flow and capable of influencing the grade of service offered to the users. We have also discussed the methods available for evaluating their call-handling capacity, and the resulting grade of service.

However, exchanges evolve and with them the teletraffic and evaluation methods. In particular, we can expect the following innovations to emerge over the next few years.

- New services on offer to the users; alongside the traditional telephone service, there will be the generation of abbreviated dialing services, three-way conferences etc. or telecomputing services (electronic directory, gateways to data transmission networks).
- New facilities provided for operation and, in particular, implementation of the common-channel signaling system and integrated services digital network (ISDN).
- Different control architectures, e.g. architectures withstanding breakdowns without immediate intervention (fault-tolerant systems), multiservice architectures and distributed structures.

These changes in the exchanges will bring about corresponding changes in traffic engineering problems. It is doubtless presumptious to predict today what tomorrow will bring. The most we can do is to review existing problems which are already at the theoretical or laboratory stage, taking into account the trends which have already been confirmed.

It can be assumed (this is only an hypothesis) that the general structure (switching network and control system) of the exchange will remain unchanged. A concurrent assumption would be the breakdown of the network and the control system into autonomous subscriber units organized in "local networks" as the dispersed architecture in Section 7.1, or rather the two solutions would coexist as the centralized structure of the transit exchanges and highly distributed structure of the local exchanges. The American Bell System network (ESS4 and ESS5) foreshadows such an evolution.

In Section 11.1, we shall examine the new problems arising from the change in the operating modes in the switching network.

A particular type of switching network with the design and construction of supercomputers has begun to be used; this is the "interconnection network" described in Section 11.2.

Finally, Section 11.3 discusses the changes in design of the control system, i.e. the introduction of distributed structures. It is probably in this field that the most important changes should be expected.

The introduction of the common-channel signaling system and the advent of ISDN will introduce a new dimension to the study of control systems.

11.1 The switching networks

There is little to expect from the technological evolution of the switching network, at least in the short term. Time-division technology will be generalized throughout telephony at the same time as the network is digitalized. However, its use will evolve with the increase in services offered to the users and with the arrival on the scene of the ISDN.

This new approach is aimed at integrating all the services (present and future) offered to the users in a single network, at least for the local section (i.e. the distribution network up to the local exchange) in a first stage. There are a large number of advantages of this.

- From the viewpoint of the user, there is a single access to handle, a single communication procedure, a unified billing system etc.
- From the viewpoint of operation, there is a single line, a single subscription to handle and a single pair of wires to be maintained between the subscriber and the exchange.
- From the viewpoint of the network, the pooling of distinct communication resources provides economic benefits. This last point remains to be proved; in any event the profit remains to be quantified. The service quality constraints and the extremely different communication technologies may possibly cancel out the anticipated profit. The technological and economic comparison of the options is an open-ended problem.

In fact, it is not certain that the services have to be totally integrated. In any event, it seems interesting in the near future to make the access to the networks common (i.e. the part of the network which goes from the subscriber to the local exchange which provide the switching).

The services involved are extremely different in nature.

- The classical telephone service (to be more precise, access to a rate of 64 kbits/s in time-division switching or to a 3400 Hz bandwidth in space-division switching) in circuit switching; the transmission speed is provided throughout the duration of the communication.
- High-quality audio services where several 64 kbits/s circuits are allocated simultaneously (e.g. transmission of high-fidelity sound).
- The data transmission services for telecomputing applications for professional or domestic use. This is then a transmission of the packet-switching type with high or low speeds; the service no longer consists of opening a circuit for an indefinite duration but of communicating a quantity of information (from 100 bits/s to 100 kbits/s depending on the application).
- Image transmission services, videotex, facsimile or even high-quality digitized television, representing bit rates up to 200 Mbits/s.
- etc.

Depending on the strategies selected, the exchange will be a switch which will direct the call to a convenient specialized network or the services will circulate on the same transport network, the switch then becoming a multiservice switch.

As far as traffic flowing in the exchange (to be more precise in the switching network of the latter) is concerned, the mixing of services will give rise to three new problems.

- Calculation of structures which mix circuits (voice) and packet (data) switching.
- Calculation of structures where the connections are of the type n channels ($n \times 64$ kbits/s) between ends, n varying from 1 (normal telephone) to 5 or 10 (high-fidelity sound and fixed images) or even more (animated images).
- Definition of a new switching function, and broadcasting (unidirectional link between one input point and several outputs).

11.1.1 Mixing of voice and data

It is not certain that all the exchanges of the network will have to transmit voice and data simultaneously. In particular, in specialized networks where the input stage of the local exchange switches to the telephone network, the telex or data transmission network, the problem will be restricted to handling the distribution network (from subscribers to local exchanges) and the input stages.

Several solutions present themselves to quite different application areas.

- Voice packetization, which is an interesting solution when there is a preponderance of data; such a scheme is now being planned for applications of the local area network type (company networks). An appreciable gain in capacity results from the compression of information; only one direction is simultaneously active, and there is suppression of "blanks" present in the voice. This is the end of the traditional switching network, the calculation of the service quality reducing to that for an exchange (or a network) with packet switching. It is necessary to define and be able to measure and calculate the service quality offered when it is no longer measured in terms of blocking but in terms of the delay experienced by each voice sample; these delays which can impair the acoustic quality of the communication are eliminatd by buffering the packets for a sufficient duration to attenuate the fluctuations in the delays which have to be evaluated.

- Juxtaposition of the services on the same supports. In a normal switching network, data packets are routed by establishing a "virtual circuit" for each data communication. If a telephone communication is involved, a "real circuit" is established which uses a fraction of the total capacity for the duration of the exchange.
- Maintenance of the "circuit-switching" technique with possibly multiplexing of data channels handled using the same technique. This is probably the simplest solution from the implementation point of view and also from the teletraffic point of view; the calculations of traditional networks can still be applied. However, the access to data channels may be made using more complex rules (e.g. reserved channels or service disciplines with delay).

Until now, the solution involving voice packetization has been studied in greater depth, especially within the framework of local area networks (for application to actual telephone-like networks, see [Boyer 87]).

The "circuit-switching" solution has given rise to few analytical studies, although its use could now be generalized. From the teletraffic point of view, this is the "$n \times 64$ kbits/s problem" (multiple connections for a single call).

11.1.2 The problem of $n \times 64$ kbits/s connections

This problem (the establishment of communications requiring n simultaneous channels) goes beyond the scope of the switching network; since the grade of service is defined from end to end in the eyes of the user, it is also necessary to solve the problem for the interexchange network (see for example [Lutton 84]).

We assume a switching network of the three-stage network type. Under the simplifying assumption of the Lee method, the probability of blocking is

$$B_1 = P(\text{no path available}) = \{p_1 + (1 - p_1)p_2\}^V \qquad (11.1)$$

where N is the number of central matrices and p_1 and p_2 are the traffic on the links of the two stages.

If a communication requires n connections, its probability of failure will be simply

$$B_n = P(\text{less than } n \text{ paths})$$

$$= \sum_{j=0}^{n-1} \{p_1 + (1-p_1)p_2\}^{N-j}\{(1-p_1)(1-p_2)\}^j \binom{N}{j} \qquad (11.2)$$

Of course we have

$$B_1 < B_2 < \ldots < B_n \qquad (11.3)$$

The extension to more general network structures turns out to be more complex (cf. eqn (4.17)).

Another stage in the solution of the problem consists in finding out what the normal teletraffic formulas become. For simplification, we take two streams of Poisson traffic requiring n_1 and n_2 servers with intensities λ_1 and λ_2, of exponential duration with means $1/\mu_1$ and $1/\mu_2$ respectively; we assume a lost-call-cleared operating mode.

The state equations can be easily written (the state space is two-dimensional (j_1, j_2) where j_i is the number of customers of type $i = 1, 2$ in service); $P(j_1, j_2)$ denotes the probability of this state.

Let R be the total number of servers. It is necessary for the constraint

$$j_1 n_1 + j_2 n_2 \leqslant R$$

to be obeyed. It defines a domain D. The normalization condition is written as

$$\sum_{j_1, j_2 \in D} P(j_1, j_2) = 1 \qquad (11.4)$$

The solution of the problem is obtained by noting the existence of the "product form"

$$P(j_1, j_2) = A P_1(j_1) p_2(j_2) \qquad (11.5)$$

where

$$P_i(n) = \frac{(\lambda_i/\mu_i)^n}{n!}$$

A is determined by the normalization equation, eqn (11.4); it is the natural extension of the Erlang formula.

The rejection probability for the customer of type i is written as

$$B_i = \sum_{j_1, j_2 \in D'} P(j_1, j_2) \qquad (11.6)$$

where D' is the domain in which

$$R - n_i < j_1 n_1 + j_2 n_2 \leqslant R$$

(i.e. the states such that it is impossible to find n_i free servers). It is possible to define efficient algorithms for calculating A (and the B_i) (see [Roberts 81]).

The extension to more than two traffic streams is immediate and here also

$$B_1 < B_2 < \ldots < B_n \tag{11.7}$$

It remains to incorporate the "multiservice Erlang formula", eqn (11.6), into the Lee method, eqn (11.2), as was done in Chapter 4.

Nevertheless a difficulty at the practical level quickly appears when these formulas are examined; eqns (11.7) and (11.3) show that the quality offered to the most demanding service will be the worst. A rapid numerical example will show the reader the large difference between $B(n=1)$ and $B(n=5)$ for example; for traffic of 10 erlangs offered to 20 circuits, we obtain, by applying eqn (11.6), $\lambda_1/\mu_1 = 9.9$ erlangs, $\lambda_2/\mu_2 = 0.1$ erlangs, $B_1 = 5 \times 10^{-3}$ and $B_5 = 7 \times 10^{-2}$.

It follows from this that the only realistic solution consists in operating the $n \times 64$ kbits/s in a different manner; for instance, it can be operated as a waiting system or with reservation. It should be pointed out that the Erlang C formula cannot be extended as eqn (11.6); there is no "product form". It therefore remains to define the following.

- The quality criteria linked to the operating modes of the offered services (with delay and with reservation).
- The economic methods for mixing the services (in this connection, rearrangement studied in Section 6.2 is a very interesting solution).
- The methods for calculating the grade of service adapted to the selected solutions.

In the short term, digital switching with multiple connections is the general solution. Techniques of the packet-switching type (e.g. the so-called "asynchronous time-division" switching) are likely to emerge when wide-band communication comes into general use (mainly in television) (see [Boyer 87]).

11.2 Interconnection networks for multiprocessor machines

The increasing needs of the users of data processing and the advent on the market of low-cost processors have caused and allowed the implementation or the design of supercomputers containing a large number of processors. Thus, mainframes have been equipped for a long time with two or three processors which share the tasks and are connected by a bus to a common memory. In 1972 the computer STARAN already contained 8192 processors.

This mode of operation where the processors act independently is becoming inadequate and these days the requirements of real-time calculation require organizations of cooperating processors (parallel processing) containing up to 2^{14} elements or more [McPease 77].

This involves, for example, calculations on giant matrices, image processing (weather applications), voice processing, solution of large differential systems with finite-element techniques (aerodynamics), multidimensional Fourier transform calculations etc.

Thus a correlation calculation on a digitized image decomposed into 200 000 points with intensity X_i, $i = 1, \ldots, 200\,000$ requires an iterative calculation, each step handling X_i and $X_i + 1$; 200 000 processors working in parallel will handle the problem in a single cycle. In [ACM 71] an overview of the existing systems and their application is given.

The interconnection of such a number of processors requires a more elaborate device than a simple bus and the designers of parallel computers have (re)discovered the virtues of networks which they have called interconnection networks.

- Connecting networks operating using circuit switching similar to those of telephony.
- Packet-switching networks in which the nodes may temporarily store information.
- Passive interconnection structures (fixed wiring providing static interconnection).

These computers operate in two main ways.

- The single-instruction multiple-data-stream machine. A control unit distributes the same instruction to all processors and they or some of them execute this instruction simultaneously on different data; the interconnection network provides the communication between the processors and the memory cells. The control unit may also execute the part of the code which cannot be executed in parallel.
- The multiple-instruction multiple-data-stream machine. Each processor may execute its own instruction and follow its own stream of instructions. One control unit may orchestrate the operation. The interconnection network will have the same function as in the single-instruction multiple-data-stream machine.

At a detailed level, important differences appear between the interconnection network and the telephone-connecting network.

- Certain networks allow the passage of data in "broadcast" mode (from a memory to several processors), i.e. sending of the lines of one matrix to all the processors for a product of matrices.
- A machine (e.g. single-instruction multiple-data-stream) must be able to function as a network of nN processors or as n machines of N processors.

- The structure of processors and memories may be considered as fixed at the design stage; the problems of modularity and growth do not exist.
- The number of operations, the number of crosspoints and the times are the critical parameters.
- Of course the cost of the interconnection network which represents a large part of the total cost must remain low.

A large variety of structures are found depending on the number of processors to be connected and the type of applications. The classification shown in Fig. 11.1 is from [Feng 81].

Static structures produce a fixed interconnection. In this case the roles of interconnection and calculation are merged, each processor controlling part of the memory. In general, a message has to pass through several intermediate nodes. Thus, Wittie suggests (Fig. 11.2) a structure called the "dual-bus hypercube" [Wittie 81].

The dynamic structures resemble telephone-switching networks. The square $n \times n$ matrix (crossbar) is limited in size and the designers have looked for economic structures. The nonblocking networks turn out to be too large (a

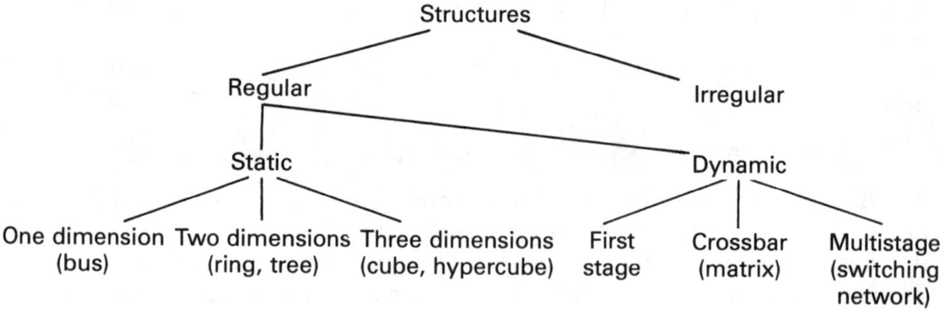

Fig. 11.1 Classification due to Feng.

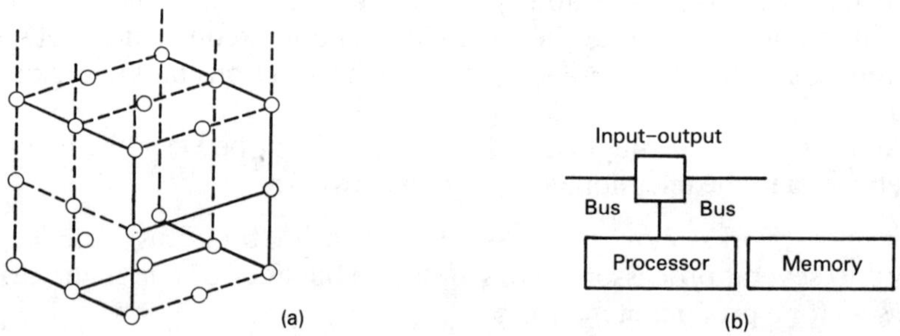

Fig. 11.2 (a) The network architecture and (b) a node of the "dual-bus hypercube" network: ——, bus. Each processor is connected to two buses.

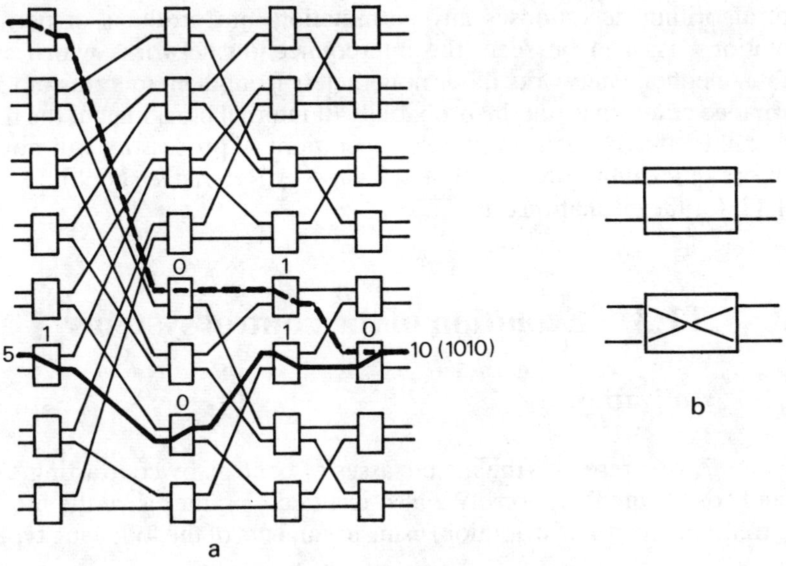

Fig. 11.3 The network in [Barnes 81]: (a) the network; (b) the two states of the elementary switch.

large number of nodes and stages and so prohibitive transit times); they are most often directed to blocking networks. Similarly, most often, centralized control has been abandoned. We describe as an example a network (similar to the STARAN network) developed for an aerodynamic simulation application [Barnes 81]. The elementary matrix is a 2×2 element for technological reasons and has two states (Fig. 11.3).

The operation is as follows: each stage i decodes the bit of rank i of the address carried by the incoming message and positions itself according to its value (1, connection to the bottom output; 0, connection to the top output). The diagram shows the example of two possible connections to the output 10 (binary coded 1010).

There is only one path to one outlet. When conflict occurs, the node blocks one of the calls which will be served later. Each operation cycle will consist of subcycles in which the calls blocked in this way will be successively routed.

The evaluation of the performances shows the limitation in the gain due to these conflicts [Barnes 81].

Other implementations use another way of resolving the conflict; the blocked calls are switched to another output and reach their destination at the end of several transits of the network [McPease 77].

The network is used as a "permutator", only realizing a fraction of all the possible permutations. For example, the network in Fig. 11.3 realizes $2^{16} = 64\,000$ permutations among the $16! = 2 \times 10^{13}$ possible permutations. The control algorithm decomposes any permutation into products of permissible permutations. As can be seen, the interconnection networks which strongly resemble telephone networks differ nonetheless from them in terms of use. The performance criterion is not the probability of internal blocking but the time lost or rather the effective gain in parallel operation; N processors will not run N times more quickly but only aN times $(0 < a < 1)$. Reference should be made to [Siegel 82] for the evaluation criteria.

11.3 Evolution of the control systems

11.3.1 Motivation

In Chapter 7, we presented the main classes of control, by contrasting "centralized" and "distributed" options. We also justified or rather gave the motivation for the trend in favor of distribution, using arguments of the following type.

- Technology (search for large capacities).
- Reliability (better use of redundancy and improved availability).
- Economic (linear growth of the cost and standardization of the system range).

The above reasons are based on the architectural viewpoint. As will be seen later (see Section 10.4), the traffic engineer has to cope with other considerations when trying to forecast the evolution, i.e. the emergence of "new services" which force the exchange to enter the ISDN era.

After reading and reflecting on the previous chapters, the reader will undoubtedly make sense of the classification of Chapter 7; (at the traffic level) centralized structures are defined by the existence of a single master which controls the functioning of the system. Decentralization consists in transferring some of the most repetitive tasks from the master to the peripheral stations while still leaving the control to it (in particular, the master retains the responsibility of allowing new calls to enter).

In a distributed architecture the peripherals become autonomous; they turn to the pooled resources (processors, data or programs) in order to solve their problems and can choose the location of these. The pooled resources are standard or specialized. Of course, actual current implementations lie somewhere between these extreme positions.

Each solution has clear-cut advantages. Centralization can claim to provide the best solution to a certain number of technical problems such as absence of conflict in the seizing of resources, optimization of intercommunication and data management. Since distribution seems an inescapable trend, it will have to cope with and solve the problems which it raises. We briefly review some of them in the following two sections. Subsequently, Section 10.4 will discuss how distributed structures can facilitate evolution towards ISDN.

11.3.2 Load sharing between units

Certain units (e.g. processors) are organized on the basis of traffic sharing and an incoming request must choose between those available. In contrast with a classical *GI*/*GI*/*R* queue where the choice does not affect the performances, the internal state of units as complex as processors must be taken into account in the choice.

The following considerations are by no means restricted to SPC exchanges and apply equally to any multiresource system. Peripheral units generate different types of task (say, with unequal length or even unequal urgency). However, the shared processors may not be equivalent as they may have different speeds and different instantaneous loads; moreover, some call-handling or maintenance tasks may be dedicated.

In this organization the simplest way of balancing processor loads is by a uniform distribution of incoming requests over the servers, without analyzing their internal state; on average, of course, the processors will thus be equally loaded, although local imbalances may occur.

Very simple models can be constructed for the job allocation mechanisms in such multiprocessor systems (Fig. 11.4).

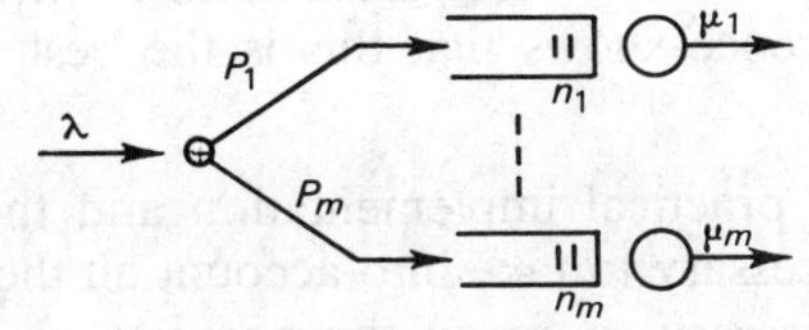

Fig. 11.4 Task allocation among servers.

A distinction should be made between random and deterministic allocation policies. Most often, a (pseudo)random rule is adopted (in fact, a revolving priority system is involved). An elaborate rule cannot be defined without taking information from the system. According to [Chow 79], we shall define nondeterministic routing as follows: the server i is selected with probability p_i. The resulting average sojourn time (each station is represented by a *M*/*M*/1 queue) is

$$T = \sum_{i=1}^{m} p_i \frac{1}{\mu_i - \lambda p_i}$$

For example, let us choose p_i proportional to μ_i; we have

$$T = \frac{m}{\sum_{j=1}^{m} \mu_j - \lambda}$$

This is an equitable choice producing the same utilization level for all servers.

- The policy of a minimum response time which is a deterministic policy. Let us assume that the system is in the state $(n_1, n_2, \ldots, n_m)$; an incoming customer is sent to the queue with a minimum ratio $(n_i+1)/\mu_i$. It is clear that in this way the incoming customer chooses the queue which seems the best choice, promising him the minimum response time.
- The policy of minimum time in the system. The preceding policy tends to optimize the waiting time of the customer. Here we have the opposite; the queue is chosen which minimizes the total service time of all the customers present. The total service time is defined as the time until the departure of the last customer present. Let $T(n_1, \ldots, n_m)$ be this time; we shall then choose i such that $T(n_1, \ldots, n_i+1, \ldots, n_m)$ is a minimum.
- The policy of maximum flow rate. If the switching mechanism has information on the arrival rates, it can implement very sophisticated policies. Thus, it can be shown that the average expected flow rate, the state being given, is:

$$TP(n_1, \ldots, n_m) = \sum_{i=1}^{m} \lambda_i \left\{ \sum_{k=1}^{n-1} \left(1 - \frac{n_i}{k}\right) \left(\frac{\mu_i}{\lambda + \mu_i}\right)^k - n_i \log_e \left(\frac{\lambda}{\lambda + \mu_i}\right) \right\}$$

The queue i can then be chosen which maximizes $TP(n_1, \ldots, n_i+1, \ldots, n_m)$. The above-mentioned reference shows that this is the best rule (it minimizes the average waiting time).

The difficulty of practical implementation and the evaluation of these policies is that it is necessary to take into account all the sources of processor activity. It is also necessary to know the parameters μ_i and to be able to "measure" the state of the system.

11.3.3 Access to common objects

One of the objectives of the multiprocessor architecture is the increase in capacity which is expected to be proportional to the number of processors. The efficiency or the power of the organization is defined as the ratio of the capacity to that of a monoprocessor machine.

The phenomena of conflict for access to common resources contradict the theoretical proportionality; the waiting in front of the connection system (bus) and the memories have a braking effect on the speed of each processor.

The number of buses can be multiplied, even calling on an interconnection structure; there still remains access to the memories where sharing cannot be completely avoided. Several solutions may be adopted in the case of conflict for access (for reading and writing or reservation during processing).

- Passive waiting for the memory.

- Abandonment, i.e. the processor placing the task in reserve in order to re-activate it later (it is perhaps another processor which will process it).

The last solution causes losses in time (saving and loading operations and the risk of abandonment of the task by time delay when there is overload; this is a natural collapse mechanism). In contrast, it allows the processor to serve another call. No preference can be given to one of the solutions, given the actual state of knowledge and from the traffic viewpoint.

In the case of waiting, a simple model allows a qualitative study of the phenomenon.

Let $1/\lambda$ be the average time (exponential distribution) between successive access operations to the memory by a processor, $1/\mu$ be the occupation time (exponential) and M be the number of processors. This is the framework of the finite-source model; a birth and death process governs the evolution (Fig. 11.5).

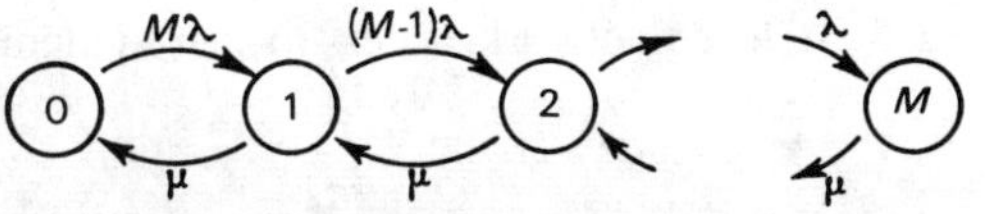

Fig. 11.5

A state is defined by the number n of waiting processors ($0 \leqslant n \leqslant M$). The solution is easily obtained:

$$P(n) = \left(\frac{\lambda}{\mu}\right)^n \frac{M!}{(M-n)!} P(0)$$

We set $a = \lambda/\mu$ and

$$P(0) = \frac{a^M/M!}{\sum_{i=0}^{M} (a^i/i!)} \quad \text{(the Erlang } B \text{ formula)}$$

The average waiting time is

$$W = \sum_k \frac{k}{\mu} P(k) = \frac{1}{\mu} [M - a\{1 - E(M, a)\}]$$

and the efficiency of the organization is given by

$$\text{efficiency} = M \frac{W + 1/\lambda + 1/\mu}{1/\lambda + 1/\mu}$$

In practice, there will be several memory blocks resulting from the partitioning of the common space. Furthermore, certain information will be duplicated. The following model generalizes the preceding model. q_i denotes the probability that an access is made with the destination of the block i (Fig. 11.6).

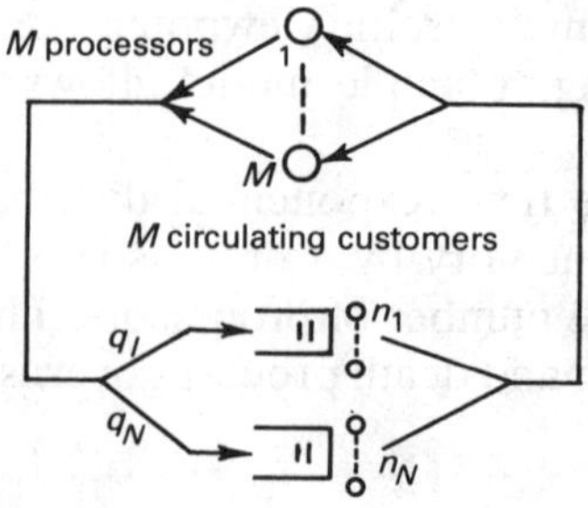

Fig. 11.6 Closed network: access to shared memories.

This is a closed queuing network. The processor time between one access and another may be assumed to have any distribution. The BCMP theorem makes it possible to calculate the state probabilities and the Little formula then gives W.

The following points should be noted.

(1) In a centralized system where the two computers operate by sharing the traffic, the same phenomenon produces the same effects; the "duplex" capacity is not double the "simplex" capacity (cf. Section 7.1).

(2) In [Sastry 75] this problem is touched on and various memory organizations are compared. The measurement parameter used is the "instruction execution rate". If it is related to that of the monoprocessor system, it is equivalent to the efficiency.

In [Marsan 82], other criteria are defined for measuring the performance and are related to each other (in particular, conflicts for the bus are studied).

These two references can be placed within the framework of a general-purpose multiprocessor computer (not specialized for switching).

11.3.4 In conclusion: centralized or distributed?

We have already pointed out the conclusive advantages of distributed systems. In contrast, we have just discussed two specific problems which they raise. Other points such as overload have been already mentioned in Chapter 10. Once again, since the traffic conditions must clarify and guide the process of choosing the architecture, the examination of these points can be of some help in the design process of modern systems.

There is always a choice between the alternatives centralized or distributed systems or, within the distributed option, between universal and specialized systems with respect to processing power, stored data, peripherals and programs. It is probable that, although the completely centralized solution is excessive, the dual entirely distributed solution is also excessive. An optimum should therefore be found in the distribution of data and in their duplication; and similarly for the software and the hardware; the question of the degree of centralization (or distribution) is posed for each "resource".

Another way of expressing the problem is to determine how the privatization of resources should be organized. Figure 11.7 taken from [Chaniac 80] briefly summarizes the points of comparison.

In addition, it is necessary to bear in mind that a choice is only optimal relative to a given state of technology.

11.4 Common-channel signaling system and integrated service digital network

The introduction of common-channel signaling will modify telecommunication networks and obviously will have some effect on the exchanges.

The final stage consists of a specialized data transmission network which as a whole acts as a "server" for the telecommunication network. At present, CCITT Standard 7, the protocol devoted to common-channel signaling, is used on a channel-associated basis; in the CEPT standard, the time slot 16 carries signaling information on 30 channels.

In terms of the relationship to switching and control functions, the effect will result from the following.

(a) New facilities resulting from the enrichment of signaling codes. Maintenance will be facilitated as well as technical management; it will be possible to devise procedures for modifying routing in real time, reacting to the various incidents which occur in the network. Similarly, the expected speed of common-channel signaling makes it possible to implement the preliminary subscriber test; the state of a distant called party is tested before commencing the connection procedure. This results in an increase in the control efficiency, the control no longer having to route the 10–15% of the ineffective traffic due to the fact that the subscriber is busy.

(b) Modification of the call flow diagrams due to the change in procedure and the new facilities mentioned in (a).

(c) The new facilities of the signaling code will be fully used in the ISDN; each "new service" will give rise to a new type of call.

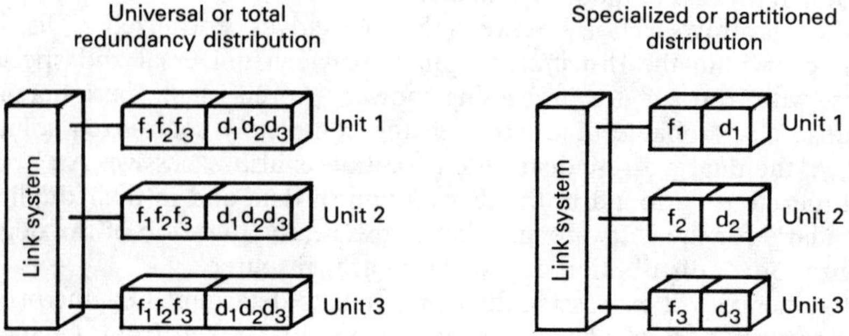

Parameter	*Each unit possesses all (or nearly all) functions and data, i.e. universal distribution*	*Each unit only has a small number of functions and data; i.e. specialized distribution*
Memory capacity	Maximum because of copies	Minimum
Availability security	Excellent (parallel *n*-plication)	Low (the units are in series from the reliability point of view)
Messages between units	Minimum speed	Maximum speed (the greater the specialization, the higher is the speed)
Total duration of processing	Minimum; once processing is accepted by a unit, it runs entirely inside this unit	Maximum, since the transfer from one function to another is effected by messages and therefore slowly
System supervision	Complex (variety of distinct functions)	Simple at the level of each unit but more complex at the level of the network which they form
Coherence of copies	Difficult to solve for global evolving data	No problem

Fig. 11.7 Universal or specialized distribution : f, function; d, data. Which should be selected? The selection criteria.

The last point brings the discussion to the ISDN. As far as ISDN functions are concerned, current standards provide subscribers with the so-called basic "$2B + D$ access"; a 144 kbits/s rate is shared between two circuit-switched B channels, 64 kbits/s each, and the packet-switched D channel (16 kbits/s) used for signaling and data transmission.

The subscriber unit has to manage the link, by analyzing signaling information and switching the corresponding connections towards specialized networks or internal servers.

It lies outside the scope of the book to describe in detail protocols and services. We are only interested in their call-handling and performance aspects. From this point of view, common-channel signaling and ISDN will give rise to the same developments.

First of all, with common-channel signaling and ISDN, new features are interpreted by new types of call or by new call flow diagrams; this amounts to adding new sequences to the call mix in Chapter 8 or modifying the task sequences of previously defined calls.

The new sequences closely resemble those previously described. The methodology in Chapters 8 and 9 directly applies to the study of their performance.

The modification may be significant, especially in terms of the holding time of signaling sequences which should decrease. However, note that channel-associated or true common-channel signaling makes no difference to the call flow diagram.

However, all these new features have to be implemented. The trend seems to be the design of specialized units, acting as servers dedicated to the new tasks. This gives a distinct advantage to distributed architectures since it is clear how easy it is to add new devices in either a universal or a dispersed distributed structure.

For instance, common-channel signaling takes place quite easily in the E10 system with the design of the common-channel signaling control unit (see [Fiche 84]). The whole system remains unchanged, except that the common-channel signaling control unit is in charge of all signaling operations. At present, this unit is a pool of processors, monitoring senders–receivers permanently linked to the PCM channels (time slot 16) via the switching network. The evolution towards true common-channel signaling is restricted inside the processors and senders in the paths.

The same analysis could be carried out for all the services. The description of System 12 is another example of the adaptation of distributed systems to the current evolution (see the references to Chapter 7).

Once again, the new units resemble the classical common-control systems; the methods and results of Chapters 8 and 9 are perfectly suited to their study.

We must note, however, that new services as well as common-channel signaling rely on packet-switching data transmission networks. Let us for example take the case of signaling. In the final state of an actual common-channel signaling network, a signaling request is sent to the dedicated unit (say the UTC in E10). A message is generated, which is sent to the data transmission network; the answer eventually comes back, and the total delay takes place in the holding time of the signaling sequence. This means that the packet networks have to be studied for their own sake, and this point goes far beyond the scope of this book.

In order to summarize the discussion in this section, the following points should be remembered.

- New services or new procedures are defined as previously in terms of their "call flow diagrams".
- New devices are possibly designed and dedicated to them. The new device may even take the form of a complete network, regarded as a "server" of call-handling tasks.
- The methods of Chapters 8–10 (deterministic or analytical approaches, and simulation) can still be used for the estimation of the grade of service.
- Packet switching adds a new dimension to the performance studies (see references).

References

The following references on the evolution of exchanges complete those of Chapters 7–10. A great deal of information on ISDN (its trial implementation, the services etc.) can be found in the *Proceedings of the International Switching Symposium*. For the $n \times 64$ kbits/s problem, see [Roberts 81], [Lutton 84] and [Fiche 85]. The evolution of exchanges opens up traffic studies to the field of data transmission, which is beyond the scope of this study ([Guilbert 85] and [Halsall 85]). For common-channel signaling, see [Craveur 80].

[Boyer 87] P. Boyer, J. Boyer, J. R. Louvion, L. Romoeuf. Time transparency evaluation of an asynchronous time division network. *Proceedings of the International Switching Symposium*, Phoenix , March 1987, Paper B12.2.

[Chaniac 80] M. Chaniac, M. Morvan, J. C. Petit, J. M. Pitie. Architecture of the control of exchanges (in French). *L'Echo des Recherches*, July 1980, pp. 41–57.

[Craveur 80] J. Craveur, B. Delosme, A. Leroux, F. Lucas, P. Lucas. Common channel signalling. *L'Echo des Recherches (English Issue)*, 1981, pp. 61–69.

[Chow 79] Yuan-Chieh Chow, W. H. Kohler. Models for dynamic load balancing in a heterogeneous multiple processor system. *IEEE Transactions on Computers*, Vol. C28, No. 5, May 1979, pp. 354–361.

[Fiche 84] G. Fiche, M. Ruoven. The E10 system: functional evolution and quality of service. *Proceedings of the International Switching Symposium*, Florence, May 1984, Paper 21A2.

[Fiche 85] G. Fiche, P. Legall , S. Ricupero. Study of blocking for multislot connections in digital link systems. *Proceedings of the 11th International Teletraffic Congress*, Tokyo, 1985, Paper 5.4B.2.

[Guilbert 85] J. F. Guilbert, C. Macchi. *Teleinformatics, Data and Computer Communications, Studies in Telecommunications,* Vol. 3, North-Holland, Amsterdam, 1985.

[Halsall 85] F. Halsall. *Introduction to Data Communications and Computer Networks,* Addison-Wesley, Reading, Massachusetts, 1985.

[Lutton 84] J. L. Lutton, J. W. Roberts. Traffic performance of multislots call routing strategies in a ISDN. *Proceedings of the International Switching Symposium,* Florence, 1984, Paper 22B6.

[Marsan 82] M. A. Marsan, M. Gerla. Markov models for multiple bus multiprocessor system. *IEEE Transactions in Computers,* Vol. C31, No. 3, March 1982, pp. 239–248.

[Sastry 75] K. V. Sastry, R. Y. Kain. On the performance of certain multiprocessor computer organization. *IEEE Transactions on Computers,* Vol. C24, No. 11, November 1975, pp. 1066–1074.

[Roberts 81] J. W. Roberts. A service system with heterogeneous requirements. *Performance of Data Communication Systems,* North-Holland, Amsterdam, 1981.

The following references are on interconnection networks.

[ACM 77] Parallel processors and processing. Association for Computing Machinery, *Computing Surveys, Special Issue,* Vol. 9, No. 1, March 1977.

[Barnes 81] G. H. Barnes, S. F. Lundstrom. Design and validation of a connection network for many-processor multiprocessor system. *Computer,* December 1981, pp. 31–41.

[Feng 81] Tse-Yun Feng. A survey of interconnection networks. *Computer,* December 1981, pp. 12–27.

[Hockney 81] R. W. Hockney, C. R. Jesshope. *Parallel Computers,* Adam Hilger, London, 1981.

[McPease 77] McPease. The indirect binary *n*-cube microprocessor array. *IEEE Transactions on Computers,* Vol. C26, No. 5, May 1977, pp. 458–473.

[Siegel 82] L. J. Siegel, H. J. Siegel, P. H. Swain. Performance measures for evaluating algorithms for SIMD machines. *IEEE Transactions on Software Engineering,* Vol. SE8, No. 4, July 1982, pp. 319–329.

[Wittie 81] L. D. Wittie. Communication structures for large networks of micro-computers. *IEEE Transactions on Computers,* Vol. C30, No. 4, April 1981, pp. 264–273.

Appendix

FORMULAS AND CHARTS

The dimensioning of an exchange (calculation of the number of units, the size of the trunk groups, memories etc.) as well as the evaluation of the performances of a given system require numerical calculations some of which in practice require the use of graphs (charts).

This appendix gives the most useful graphs and the simplest formulas and algorithms.

The problem reduces to the study of the elementary queue $GI/GI/R/N$ (cf. Section 2.3). In practice, the simplest results which can be used concern the following cases

- $N = R$ (loss system) with Poisson arrivals or Engset traffic (see Sections A.1.1 and A.1.2).
- Exponential service time, Poisson arrivals (see Section 2.9 and A.2.1); this is the $M/M/R$ queue (or $M/M/R/N$).
- R and N infinite ("pure delay"); the previous results are then valid irrespective of the service distribution.

For the $M/G/R$ system, some approximations (rather efficient) are presented in Section A.2.3. Even the $M/D/R$ system (see Section A.2.2) does not give rise to simple treatment.

A.1 Loss system

The system is described in Sections 2.6 and 2.7, the notation of which will be used.

A.1.1 Infinite number of sources

The resultant arrival flow obeys the Poisson distribution with a rate λ:

$$P(k \text{ arrivals in } t) = \frac{(\lambda t)^k}{k!} e^{-\lambda t}$$

The service distribution can have any mean T; there are R servers. The offered traffic is denoted as A:

$$A = \lambda T$$

The probability of finding n busy servers (at any time or at the arrival of a customer) is

$$P(n) = \frac{A^n/n!}{1 + A + A^2/2! + \ldots + A^R/R!}$$

In particular, the rejection probability is

$$E(R, A) = \frac{A^R/R!}{\sum_{i \leqslant R} (A^i/i!)} \tag{A.1}$$

This is the "Erlang B formula". It can be easily programed using the recursion

$$E(0, A) = 1$$

$$\frac{1}{E(n, A)} = 1 + \frac{n}{A}\frac{1}{E(n-1, A)} \tag{A.2}$$

The resulting algorithm has excellent performances executed on an ordinary computer. On a pocket calculator, it is rather slow and becomes difficult to use as soon as $R > 50$. In most cases, the following approximation can be used: replace the denominator of eqn (A.1) by e^A and replace the factorial by the Stirling formula:

$$E(R, A) \approx \left(\frac{A}{R}\right)^R \frac{e^R e^{-A}}{(2\pi R)^{1/2}} = \frac{1}{(2\pi R)^{1/2}} e^{R - A + R \log_e (A/R)} \tag{A.3}$$

This formula gives excellent results as soon as E is low (the relative accuracy better than 5% as soon as $E < 1\%$); thus, we have the following.

- For $R = 100$ and $A = 80$, eqn (A.1) gives $E = 3.99 \times 10^{-3}$; the calculation time is close to 1 min on the HP 41C.
- Equation (A.3) gives $E = 3 \times 94.10^{-3}$ (with a calculation time of the order of 1 s).

The diagram in Fig. A.1 shows the domain of validity of the approximation in eqn (A.3) as a function of A and R.

Other approximate formulas have been suggested with less spectacular efficiency (see [Jagerman 74] and [Farmer 78]).

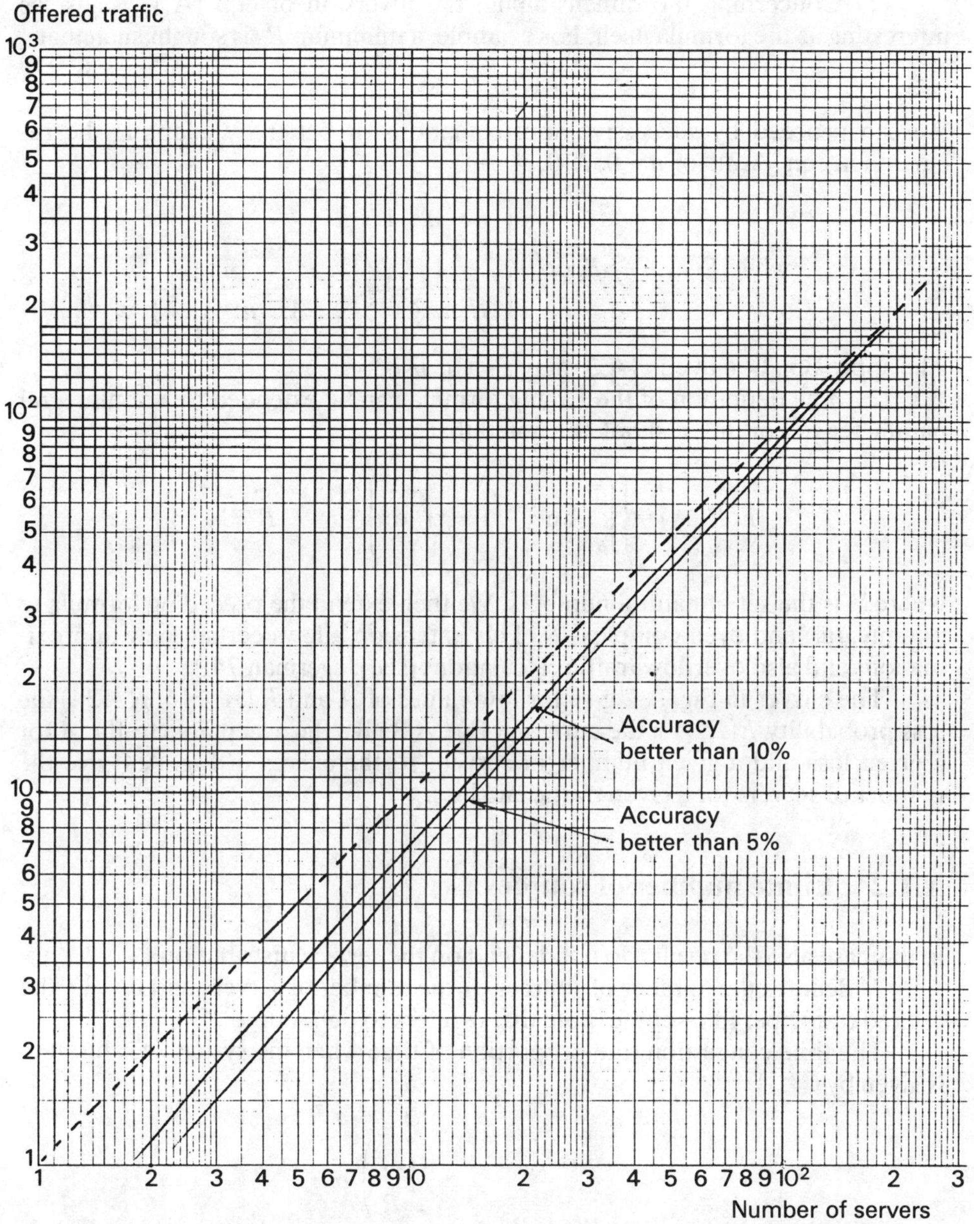

Fig. A.1 Validity of the approximation of the Erlang formula, eqn (A.3).

The following points should be noted.

(1) Concerning the dimensioning, the inversion of eqn (A.1) is just as interesting as the formula itself. For example, a minimum R_0 is sought such that

$$E(R_0, A) < B$$

We can proceed by successive approximations or better by adapting the recursion in eqn (A.2); for $n=0$, $E(0, A)=1$

$$\text{while } E(n, A) > B, \text{ do } \begin{cases} n \leftarrow n+1 \\ \dfrac{1}{E(n, A)} = 1 + \dfrac{n}{AE(n-1, A)} \end{cases}$$

The approximation in eqn (A.3) may also be used.

(2) The definition of the Erlang formula can be extended to a nonintegral parameter R. For this, it should be noted that

$$E(R, A) = \left\{ \int_0^\infty A\, e^{-Ax} (1+x)^R \, dx \right\}^{-1} \qquad R \in \mathbb{N}$$

where $\mathbb{N}$ is the set of natural numbers. We then extend the preceding formula to nonintegral (and even complex) R. This extension allows certain new approximations and also overflow traffic calculations (see [Jagerman 74]).

The curves in Fig. A.2 give for low values of A and R ($A \leqslant 16$; $R \leqslant 24$) the loss probability $E(R, A)$; the curves in Fig. A.3 give the acceptable traffic A for a given loss and a given number of servers R; the curves in Fig. A.4 give the number of servers for a given traffic and loss.

A.1.2 Finite number of sources

In this case also the results do not depend on the service distribution.

M denotes the number of sources, R the number of servers and $\alpha = \lambda T$ the offered traffic per idle source (note that α refers to one source).

The time congestion (the proportion of time when the system is saturated) is given by

$$B_T = B(M, R, \alpha) = \frac{\dbinom{M}{R} \alpha^R}{\sum\limits_{j \leqslant R} \dbinom{M}{j} \alpha^j} \tag{A.4}$$

The call congestion (probability of rejection) is given by $B(M-1, R, \alpha)$.

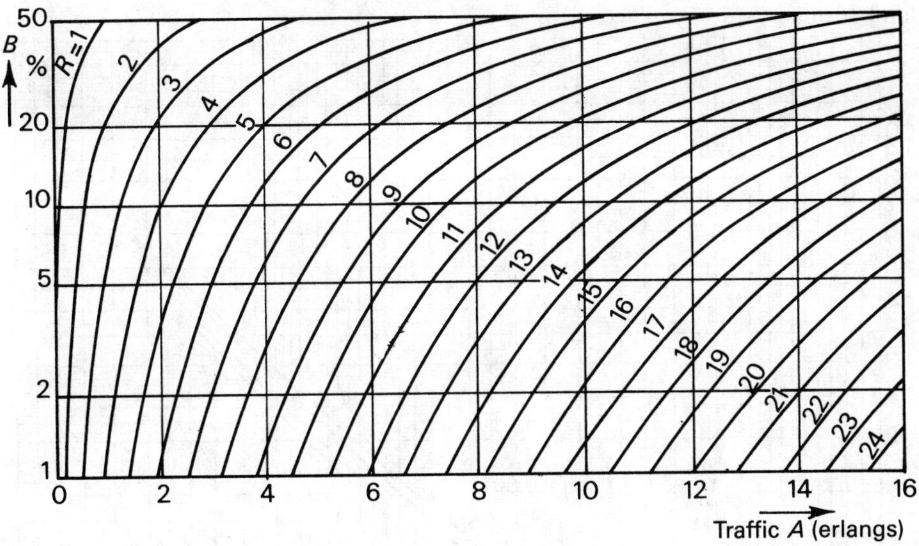

Fig. A.2 Loss as a function of traffic and a number of circuits.

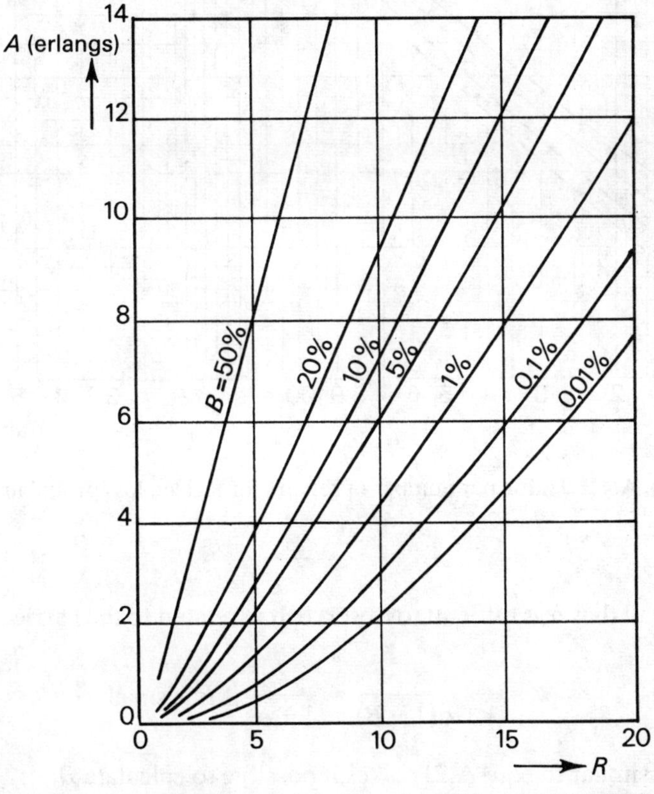

Fig. A.3 Authorized traffic as a function of the permissible loss.

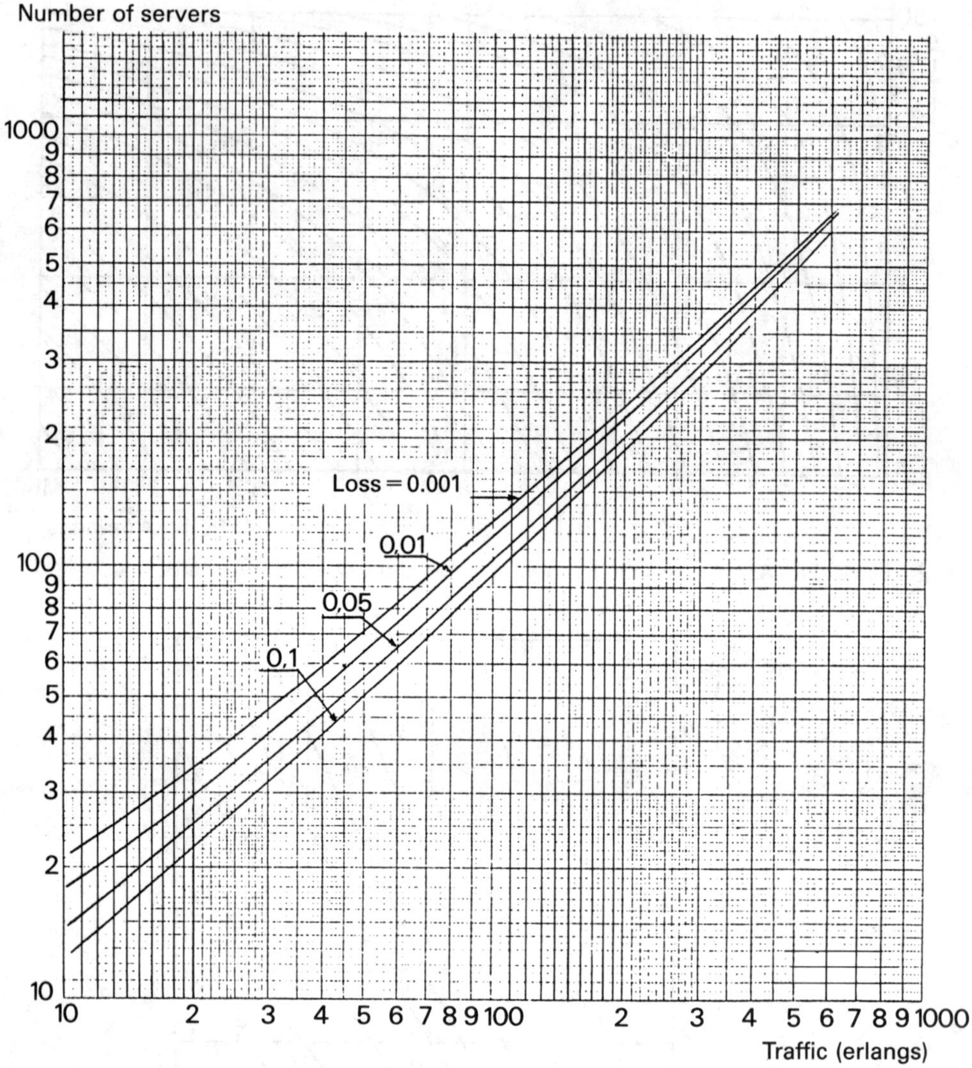

Fig. A.4 Traffic per number of servers for a given loss probability.

We recall that α is difficult to observe. It is related to the carried traffic by

$$a = \frac{\alpha}{1 + \alpha(1 - B)} \approx \frac{\alpha}{1 + \alpha} \text{ if } B \text{ is small}$$

A recursion similar to eqn (A.2) makes it possible to calculate B:

$$B(M, 0, \alpha) = 1$$

$$\frac{1}{B(M, j, \alpha)} = 1 + \frac{j}{\alpha(M - j + 1)} \frac{1}{B(M, j - 1, \alpha)} \tag{A.5}$$

When M is sufficiently large compared with R, we obtain the Erlang formula once more. Nevertheless, it requires a large M/R ratio for the Erlang formula to give better than an order-of-magnitude value. We use the Erlang formula with offered traffic $A = M\alpha/(1 + \alpha)$ (for low loss).

An example is given in Table A.1.

Table A.1

M	α	R	B	
			Engset formula	*Erlang formula*
40	0.05	5	0.0295	0.0313
40	0.1	10	0.0019	0.003
40	0.15	10	0.0185	0.0224

When M/R is of the order of 1–2, the Bernoulli formula may be used. Unfortunately it also contains three parameters and is therefore as difficult to handle as eqn (A.4).

Figure A.5 gives the total traffic as a function of M and R for a loss probability of 1%.

A.2 System with delay

Contrary to loss systems (and Poisson arrival systems), the service distribution has a strong influence on the queue characteristics. The service discipline is also involved.

The characteristics affecting the dimensioning are as follows.

- The maximum acceptable traffic.
- The probability of delay.
- The quantiles of the waiting time (at least the mean and the variance).
- The distribution of the number in the queue.

We shall limit our review to Poisson arrival systems. As previously, R denotes the number of servers, λ as the arrival rate and T the average duration of service. The offered traffic, equal here to the carried traffic, is $A = \lambda T$.

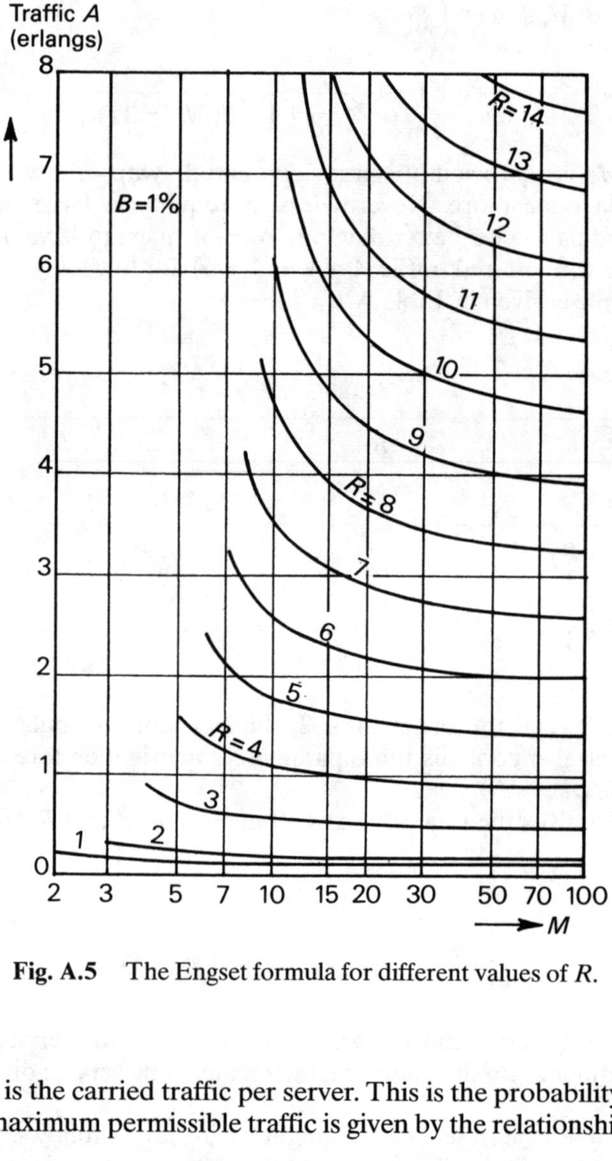

Fig. A.5 The Engset formula for different values of R.

$\rho = A/R$ is the carried traffic per server. This is the probability that a server is busy. The maximum permissible traffic is given by the relationship

$$\rho = \frac{\lambda T}{R} < 1 \tag{A.6}$$

A.2.1 Exponential service distribution (*M/M/R* queue)

The state probabilities are known (see Section 2.9). The waiting probability (the "Erlang C" formula) is

$$P(\text{waiting}) = E_2(R, A) = \frac{A^R/R!}{\rho A^R/R! + (1-\rho)\sum_{i \leq R}(A^i/i!)} \tag{A.7}$$

It is given in Fig. A.6. It is calculated via the Erlang B formula

$$\frac{1}{E_2(R, A)} = \rho + (1-\rho)\frac{1}{E(r, A)}$$

The distribution of the waiting time will depend on the service discipline. The average waiting time does not depend on it:

$$EW = \text{average waiting time (not including the service)}$$

$$= E_2(R, A)\frac{T}{R-A} \tag{A.8}$$

The average sojourn time will simply be $EW + T$.

Figure A.7 gives the normalized waiting time (i.e. EW/T) for $R < 100$.

As an example, let us consider offered traffic of 8 erlangs to $R = 10$ servers. Figure A.7 gives, for $R = 10$ and $\rho = A/R = 0.8$, $EW/T = 0.2$. Figure A.6 gives $E_2(10, 8) = 0.4$. Equation (A.8) is well verified.

The probability distribution has a certain importance; we have seen that the dimensioning was often made for a quantile $P(\text{waiting time} > t_0) = p$.

We write

$$P(\text{waiting time} > t) = E_2(R, A)F(t)$$

with $F(0) = 1$, since obviously $1 - E_2$ gives the probability of no waiting.

For service in the order of arrivals,

$$F(t) = \mathrm{e}^{-t(R-A)/T} \tag{A.9}$$

For a random service, Fig. A.8 gives the curve F as a function of the parameter ρ (note that the curve is not graduated in values of t but in values of Rt/T).

As an example, for $R = 10$ servers and $A = 5$ erlangs, the probability of a waiting time more than $2T$ is ($\rho = 0.5$; $Rt/T = 20$)

$$P(>2T) = E_2(10, 5)F(20) = (3.5 \times 10^{-2}) \times (1.7 \times 10^{-3})$$

As a comparison between service disciplines; for a traffic of 8 erlangs offered to $R = 10$ servers, what is the time that 1 chance out of 100 is reached or exceeded?

$$P(t_0) = E_2(10, 8)F(t_0) = 0.01 F(t_0) = \frac{0.001}{0.4} = 0.025$$

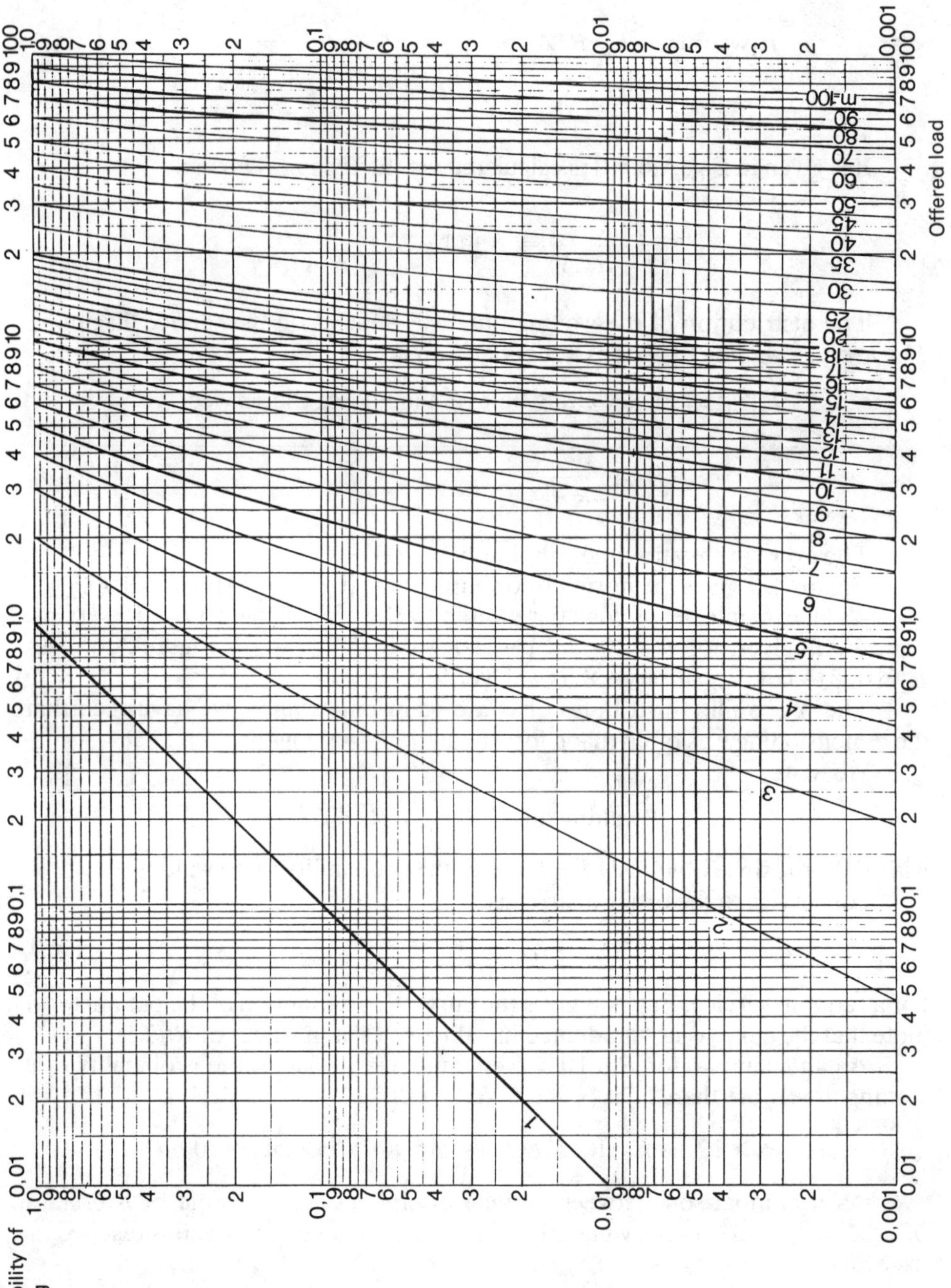

Fig. A.6 The Erlang formula with delay.

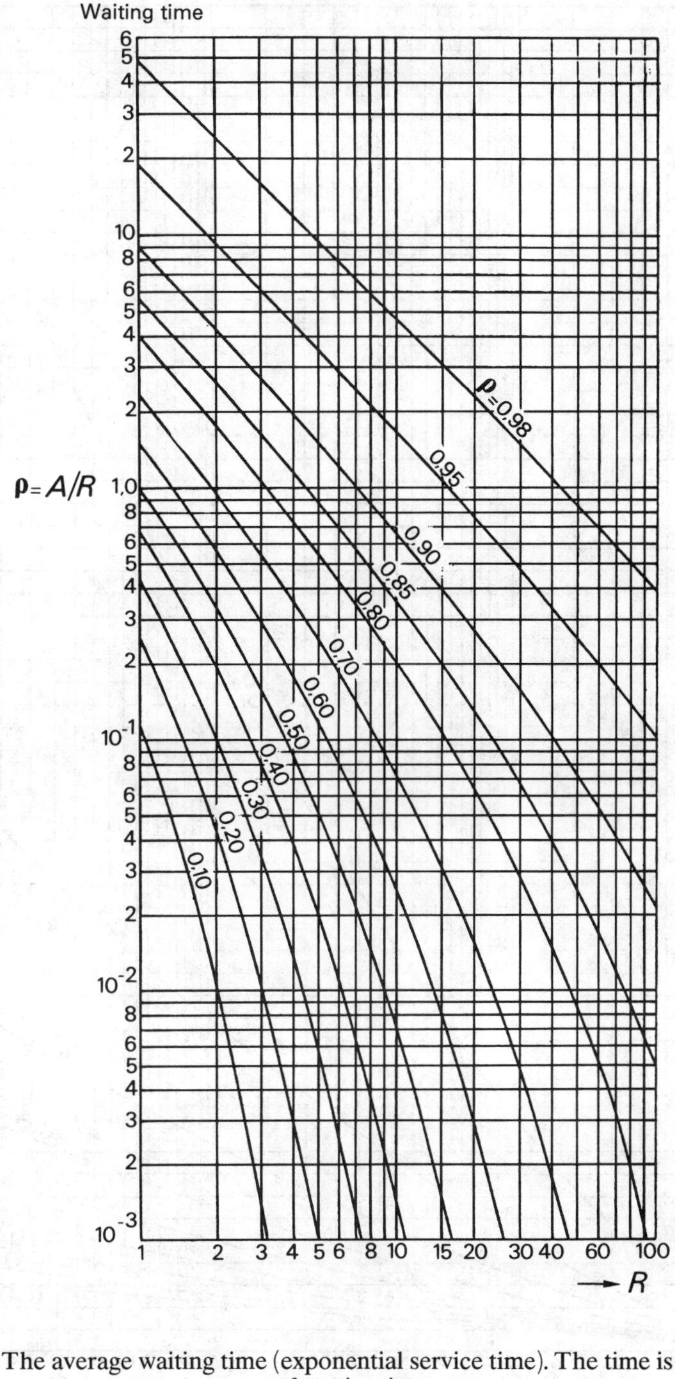

Fig. A.7 The average waiting time (exponential service time). The time is given in units of service time.

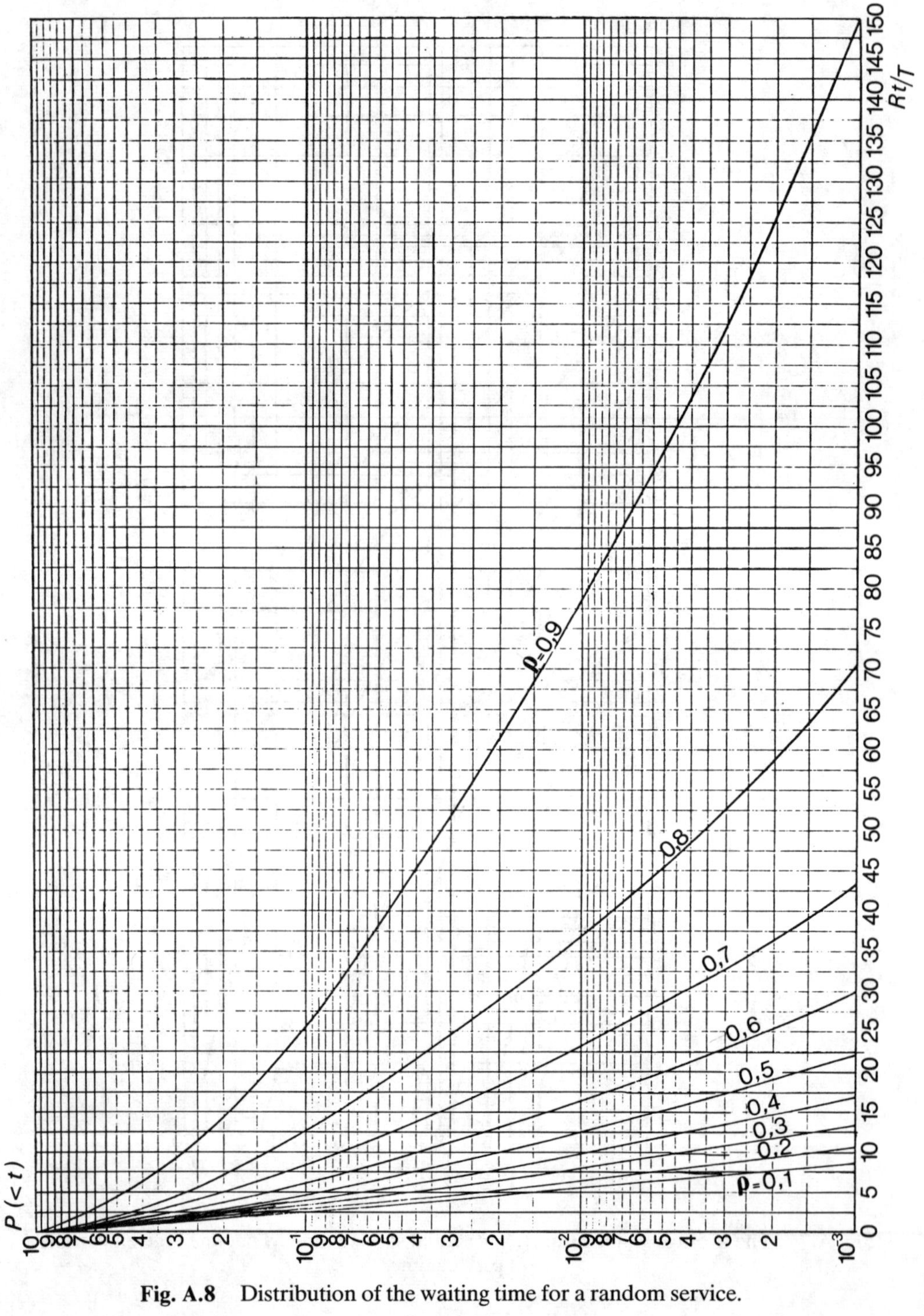

Fig. A.8 Distribution of the waiting time for a random service.

For service in the order of arrivals,

$$\frac{t_0}{T} = \frac{\log_e\{F(t_0)\}}{R-A} = 1.9$$

For random service the curve in Fig. A.8 gives $t_0/T=2.7$, for $F(x)=0.025$, $\rho=0.8$ and $x=Rt_0/T=27$.

The 99% quantile of waiting times will be larger for random service than for service in the order of arrivals. This is not always the case (since the means are identical).

A.2.2 Constant service distribution (M/D/R queue)

The use of graphs is strictly necessary here; no closed analytical form is known for the quantities of interest (see [Syski 86]).

Figure A.9 gives the probability of waiting as a function of A and R. In an approximate manner, it can be stated that

$$P(\text{waiting } M/D/R) \approx E_2(R, A)$$

As an example, for $A=4$ erlangs, $R=10$ and an $M/D/R$ queue of 8×10^{-3}, $E_2(R, A)=8.5\times10^{-3}$.

As far as waiting is concerned, Fig. A.10 gives the average time of calls which are waiting. The average waiting time $E(W)$ can be deduced from this:

$$E(W) = \text{average time of calls which are waiting} * P(\text{waiting}).$$

Thus, for $A=8$ erlangs and $R=10$, the average time of calls which are waiting is $0.3T$. The waiting probability is 0.4 and the average waiting time is then $0.12T$ (a comparison should be made with eqn (A.9) which gives $0.2T$).

A.2.3 The general case: approximation for the *M/G/R* queue

There is no simple result available and in practice it is necessary to resort to approximations. Out of the large number of suggestions in the literature, we shall look at one [Hokstad 78].

We have already noted the agreement between the waiting probabilities of the $M/M/R$ and $M/D/R$ systems. We shall extend this statement by the following approximation for the distribution of the number of customers:

$$P(n) = P(0)\frac{A^n}{n!} \qquad n=1, 2, \dots, R-1$$

$$P(0) = \left\{\sum_k \frac{A^k}{k!} + \frac{A^R}{(1-\rho)R!}\right\}^{-1} \tag{A.10}$$

where $\rho = A/R$.

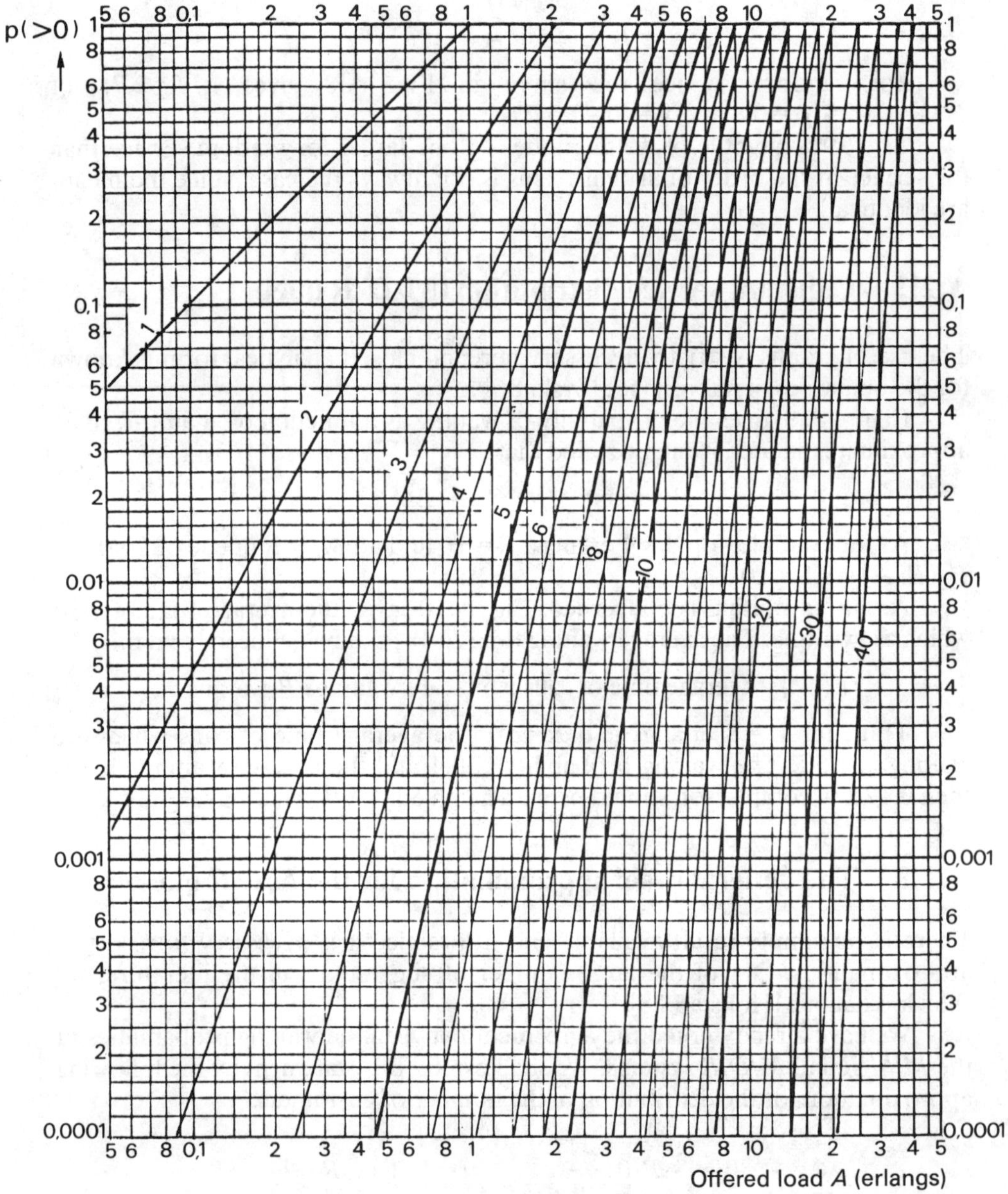

Fig. A.9 $M/D/R$ queue: waiting probability.

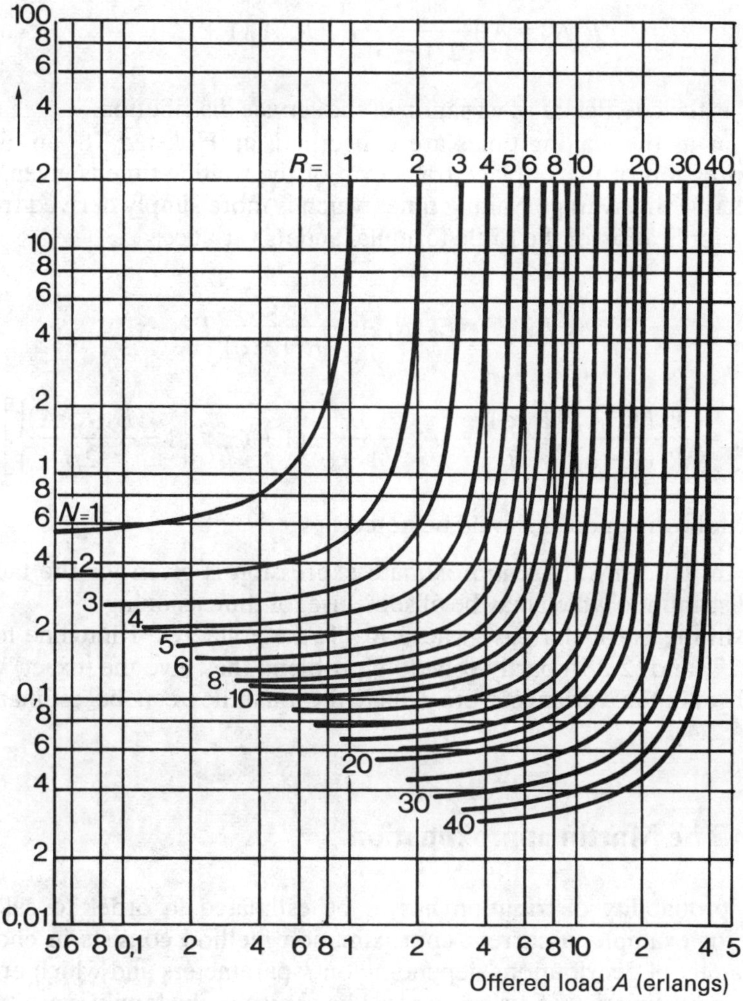

Fig. A.10 $M/D/R$ queue: average waiting time of customers actually waiting (in units of service time).

The waiting probability is therefore

$$P(\text{waiting}) = 1 - \sum_{j \leqslant R-1} P(j) = \frac{\rho}{1-\rho} P(R-1) = E_2(R, A) \qquad \text{(A.11)}$$

The approximate distribution is more difficult to write for $n \geqslant R$ (see [Hokstad 78] and also [Hoorn 82]). The average number of customers in the system is

$$E(N) = A + \frac{\rho}{2(1-\rho)} E_2(R, A)(1 + C_b^2) \qquad \text{(A.12)}$$

where C_b is the coefficient of variation of the service distribution.

As far as the waiting times are concerned, in [Hokstad 78] an (approximate) expression of the Laplace transform of the waiting time is given. This is used to derive the average waiting time (which is more simply derived from eqn (A.12) by application of the Little formula) and its variance:

$$E(W) = E_2(R, A) \frac{1 + C_b^2}{2R(1-\rho)} T$$

$$\text{(A.13)}$$

$$\sigma_w^2 = \frac{E_2(R, A)}{R^2(1-\rho)} \left[\frac{E(x^3)}{3T} + T^2 \frac{\rho}{2(1-\rho)} (1 + C_b^2)^2 \left\{ 1 - \frac{E_2(R, A)}{2\rho} \right\} \right]$$

The following points should be noted.

(1) In [Hoorn 82] an approximate expression is given for the tail of the waiting distribution which may be of some use for dimensioning.

(2) In the particular case where $R = 1$, the Pollaczek–Kintchine formulas (eqns (2.27) and (2.28)) partly solve the problem; they give the (exact) value of the mean and the variance from which the quantiles can be estimated (see Section A.2.4).

A.2.4 The Martin approximation

When a probability distribution has to be estimated in order to calculate a quantile for example, a current approximation method consists in choosing a known family of distributions depending on k parameters and which are determined by adjusting the k first moments. The choice of the family is arbitrary and Martin has observed in an experimental manner that a large number of traffic phenomena produce distributions having the general form of "gamma functions" characterized by two parameters (and therefore with rapid adjustment):

$$F(t) = P(\text{time} \leqslant t) = \frac{\int_0^x y^{C-1} e^{-y} \, dy}{\int_0^\infty y^{C-1} e^{-y} \, dy}$$

where

$$x = \frac{Ct}{E(t)}$$

It depends on two parameters, the mean $E(t)$ and the square C of the coefficient of variation given by

$$C = \left\{ \frac{E(t)}{\sigma_t} \right\}^2$$

In [Martin 72] a very complete table is given of the functions for a large range of values of C. It is summarized in the graph in Fig. A.11.

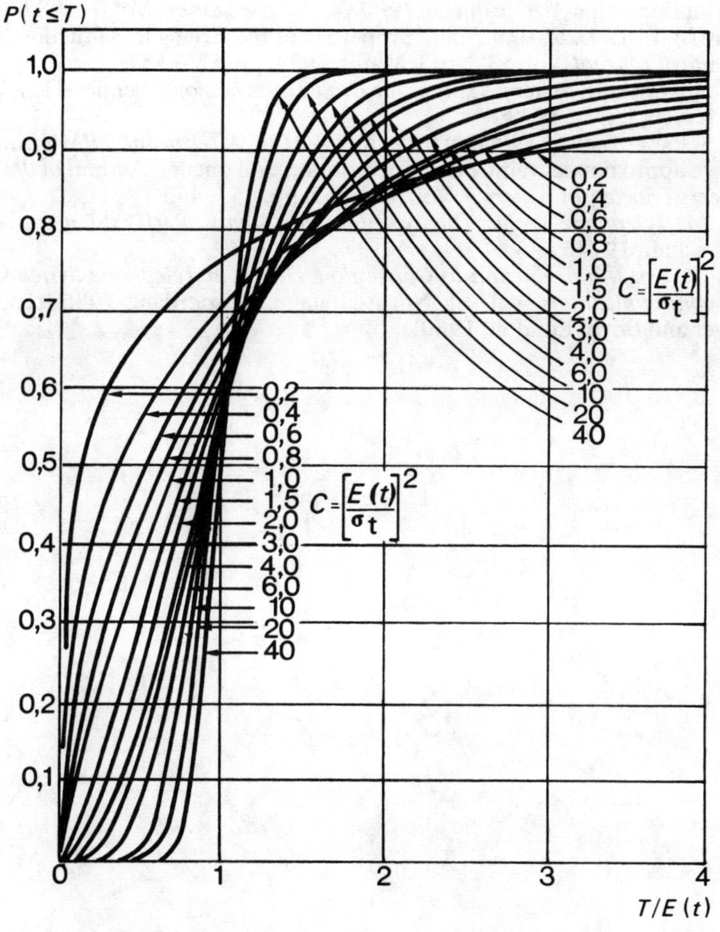

Fig. A.11 Incomplete gamma function.

References

Some of the tables and charts available for traffic engineering are listed below. A number of papers have been devoted to the efficient evaluation of traffic formulas; the main ones follow:

[Farmer 78] R. F. FARMER, I. KAUFMAN. On the numerical evaluation of some basic traffic formulae. *Networks,* Vol. 8, 1978, pp. 153–186.

[Hillier 81] F. S. HILLIER, O. S. YU. *Queueing Tables and Graphs,* North-Holland, Amsterdam, 1981.

[Hokstad 78] P. HOKSTAD. Approximations for the $M/G/m$ queue. *Operations Research,* Vol. 26, No. 3, 1978.

[Hoorn 82] M. H. VAN HOORN, H. C. TIJMS. Approximations for the waiting time distribution of the $M/G/c$ queue. *Performance Evaluation*, Vol. 2, 1982, pp. 22–28.

[Jagerman 74] D. L. JAGERMAN. Some properties of the Erlang loss function. *Bell System Technical Journal,* Vol. 53, No. 3, March 1974, pp. 525–551.

[Martin 72] J. MARTIN. *System Analysis for Data Transmission,* Prentice-Hall, Englewood Cliffs, New Jersey, 1972.

[Page 82] S. PAGE. Tables of waiting times for $M/M/m$, $M/D/m$ and $D/M/m$, and their use to give approximate waiting times in more general queues. *Journal of the Operations Research Society of America*, Vol. 33, 1982, pp. 453–473.

[Siemens 74] *Telephone Traffic Theory Tables and Charts,* Part 1, Siemens Aktiengsellschaft, 1974.

[Syski 86] R. SYSKI. *Introduction to Congestion Theory in Telephone Systems, Studies in Telecommunications,* Vol. 14, North-Holland, Amsterdam, 1986 (reprinted from Oliver and Boyd, London, 1960).

INDEX

Italicized page numbers indicate the principal reference.